ROBOTICS

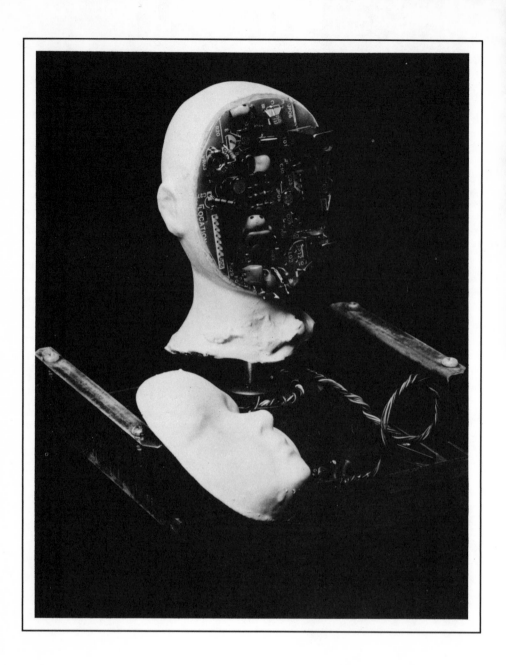

R·O·B·O·T·I·C·S

Edited by Marvin Minsky

Omni Editorial Consultant: Douglas Colligan

Photo Editor: Robert Malone

An Omni Press Book
Anchor Press/Doubleday
Garden City, New York
1985

Library of Congress Cataloging in Publication Data
Main entry under title:

Robotics.
  "An Omni Press book."
  Includes index.
  1. Robotics.   I. Minsky, Marvin Lee, 1927–
TJ211.R557   1985   629.8'92   84-24390
ISBN: 0-385-19414-5

Acknowledgments

Robotics never would have been realized without James Raimes, the book's editor at Doubleday, whose active interest in the project helped make it the special book it is. Additional thanks and credit to Bob Weil, the head of *Omni*'s book division, whose sure hand helped shape the book in its formative stage; and to Dick Teresi, the former editor of *Omni*, who also helped to launch the project.

Special thanks are due to the distinguished contributors who somehow managed to produce wonderful chapters *on time* in spite of incredibly cramped schedules: to T. A. Heppenheimer for his deft (and rapid) handling of one of the key chapters in the book; to Phil Agre for his insightful and wonderfully readable treatment of a difficult subject; to Thomas Binford for his patience and his willingness to participate in this innovative project; to Hans Moravec for taking time out from a hectic research regimen to share his very special insights; to Robert Freitas for a dazzlingly thorough portrayal of our quasi-cybernetic future; to Joseph Engelberger for his valued and original comments on an aspect of robotics that affects us all; to Richard Wolkomir for, as usual, a lively and entertaining job; to Robert U. Ayres for a thorough analysis of a complex and problematical question; to Robert Sheckley, who dazzled us all with his charmingly cynical fictional forays into the unknown; and finally to Bob Malone for the keen eye for style he brought to his job as photo editor.

Thanks also go to Mary Katherine Kneip for her fine copy editing, to Sharen Egana for taking on such a complex design project and doing it so intelligently, and to Nancy Lucas for transcribing above and beyond the call of duty. Deserving of special mention are Peter Tyson for his thoroughness and efficiency in following up on the myriad details, Bev Nerenberg, Babs Lefrak, Kevin McKinney, Cathy Spencer, Marcia Potash, and Claudia Dowling, all of *Omni*, as well as Betty Dexter of

MIT for her efficiency in helping to coordinate so many of those important early meetings.

And very special thanks to Kathy Keeton and Bob Guccione, whose interest in robotics was an inspiration to us all.

Marvin Minsky,
Massachusetts Institute of Technology

Douglas Colligan,
Omni

Contents

Contributors

Philip E. Agre is a graduate student at the Artificial Intelligence Laboratory of the Massachusetts Institute of Technology. His research has been on computational theories of personality.

Robert U. Ayres is a Professor in the Department of Engineering and Public Policy at Carnegie-Mellon University. An experienced technological forecaster and former member of the Hudson Institute, he has had a longtime fascination with the economic and social implications of technological change. He has written eight books, including *Robotics Applications and Social Implications*.

Thomas O. Binford is a professor of Computer Science at Stanford University's Department of Computer Science. Considered one of the world's leading experts in computer vision, Binford has been working on devising sophisticated vision systems for robots for almost twenty years.

Joseph F. Engelberger is the founder and first president of the industrial robot manufacturer Unimation Inc. Engelberger is given credit for being one of the creators of the industrial robot industry. He serves on the board of a number of high-technology companies and, in addition to Unimation, has founded two other successful high-tech firms.

Robert A. Freitas is a science writer/researcher who has participated as a researcher in the NASA study *Advanced Automation for Space Missions* and as the editor of another NASA study, *Autonomy and the Human Element in Space*, which examined the elements of interaction between humans and machines aboard the projected space station. An experienced astronomer and an expert in international and space law, Freitas is the director of the Xenology Research Institute, which is a leading proponent of SETI (the Search for Extraterrestrial Intelligence).

T. A. Heppenheimer has a Ph.D. in aerospace engineering and is a former research fellow of the California Institute of Technology and the Max Planck Institute for Nuclear Physics. A prolific writer and frequent contributor to *Omni* magazine, he is the author of four books and countless articles on the subject of advanced science and technology.

Marvin Minsky is Donner Professor of Science in the Department of Computer Science and Electrical Engineering at the Massachusetts Institute of Technology. He is a founder of MIT's Artificial Intelligence Laboratory and is one of the founding fathers of Artificial Intelligence (AI), the discipline that attempts to duplicate by machine the intelligent behavior of the mind. A seminal philosopher and writer on AI, Minsky is a former president of the American Association for Artificial Intelligence, the leading professional group for AI researchers and students in this country.

Hans Moravec is a research scientist at the Robotics Institute of Carnegie-Mellon University. His long-term interest is research into the use of intelligent robots in outer space. He has also been pursuing research in computer vision, supercomputers, robot manipulators, and mobile robots, an area in which he is acknowledged to be one of the preeminent experts. He has published a book on mobile robots and directs a laboratory that has built four mobile robots designed for indoor and outdoor adventures.

Robert Sheckley is a science-fiction writer best known for his books *The Robot Who Looked Like Me* and *The Seventh Victim* (which became the immensely popular film *The Tenth Victim*), as well as dozens of other works considered to be classics of science fiction. Sheckley has worked closely with Marvin Minsky in the past and recently served as writer-in-residence at MIT's Artificial Intelligence Laboratory.

Richard Wolkomir is a prolific science journalist whose work regularly appears in *Omni, Smithsonian, Reader's Digest,* and *McCall's.* He has written about all aspects of science and technology from particle accelerators to microbiology, as well as domestic robots.

ROBOTICS

Introduction
Marvin Minsky

When I was first approached with the task of doing this book I thought of several purposes for it. It could help us to understand what is happening, right now, in the new technologies already confronting us. It could help explain what is going on in research now and prepare us to face the next wave of developments in the future. And, most important, it could help us understand how our own minds work.

This book is about artificial intelligence, commonly called AI, which is an inquiry not just into computers and new kinds of machines but also into the nature of intelligence itself—a subject no one understands very well. Why not? Perhaps partly because no one's ever had a chance to look at any other kind of intelligence except our human kind! Why is this important? Because it is very hard to understand anything until you have other things to compare with it. For it is only when we examine many different examples that we can begin to sort out the important general principles from the myriad of interesting but often unimportant and accidental facts about individual instances. Thus, biologists have found out almost everything they know by making comparative studies of different kinds of animals, plants, bacteria, and viruses. Similarly, we have learned much about our own culture by studying the cultures of other nations, tribes, and communities, and we have learned a lot about our own language through comparing it to other peoples' languages and literatures. But we have never been able to study "comparative intelligence" very well—because there simply are no other animal species whose intelligence is really comparable to ours. True, we can study the way children think, which has taught us a lot about how our minds work; yet, the young human is too much like the adult. We can study chimpanzees, who are also much like us, and we can study elephants, dolphins, and dogs, whose minds are, to be sure, less like our own. But we cannot learn quite so much from them, and even less from the other animals, because they just don't have enough of what we regard as intelligence.

At last we are about to have a chance to meet some really alien minds! Our first encounters will not be in the forms of highly evolved intelligences from other planets (no one can predict when that will happen). It seems much more likely now that our first encounters with alien minds will be with ones that we ourselves have built—our very own AI machines. And, to judge by what their ancestors have been like so far, we can be confident that they'll be alien enough to give us ample food for comparative thought!

The only trouble is that we don't know how soon we're going to be able to meet them, because we don't know how to make machines intelligent! What, exactly, is the problem? In my experience, most people outside the field of AI think they already know what it is: It has something to do with the mysteries of Inspiration, Creativity, Intuition, Originality, and Emotion. "Machines," they say, "can do only what we, their human programmers, tell them to. So of course our machines can't really be intelligent: We simply do not know (and likely never will) how to tell them just what it is that our Shakespeares, Einsteins, and Beethovens do."

However, if there's one thing we have surely learned in research on artificial intelligence, it is that this is not the best way to describe the problem. Of course it will seem to be a mystery how a Beethoven writes a symphony—until one has a good idea of how to harmonize a simple tune. So I hope that this book can do two things. On one hand, it should help dispel the notion that it is hopeless to try to understand ourselves. On the other, it should encourage us to be less envious of our intellectual heroes by helping us to gain more respect for the ordinary things we all take for granted: In AI research, we are learning how amazingly complicated are the simple things we do in everyday life.

When we admire the outstanding performances of our greatest thinkers, athletes, and the like, the trouble is that we look right past the wonderful things we—all of us—do whenever we walk, talk, see, reason, and plan. It is not necessary to start AI research by trying to write computer programs that will paint great paintings or write brilliant plays. At this stage, we can learn more by trying to develop programs that can distinguish a dog from a cat by sight or carry on a simple, childish conversation.

When AI researchers first started to attempt to make computers duplicate human abilities, we encountered a curious paradox. It wasn't really very long before we had computers doing things like playing games of chess, proving theorems that puzzled mathematicians, and designing circuits that were hard for engineers to figure out. These machines—really just regular computers with programs—were very impressive. Today, their descendants are often called "expert systems," and we are finding new and useful applications for them every day. The trouble with them is that they all are too narrowly specialized; they work only within the contexts or environments they are designed for. When you try to use them for anything else, then they show few signs of having any ordinary common sense.

For example, in the early stages of designing a robot to build a tower of children's blocks, one program tried to build a tower by starting from the top—so every time it placed a block, the block would fall, and the machine had to start over again.

Why is it so hard to give machines common sense? To answer that, we must have some idea of what that is. Literally, common sense should mean "the things everybody knows." But most of that consists of things we take for granted, things so "natural and obvious" that it is hard for us to see what they are—even though they are involved in almost everything we do. For example, everybody knows that you can put a thing in your pocket—but only if it is not too large or too delicate, if it doesn't belong to someone else, or if it doesn't bite. Everybody knows that if you want to change a light bulb you can stand on a chair—but not if it's a rocking chair or a flimsy antique, or if it is a thousand miles away. Thus, one feature of commonsense knowledge is that, although at first it seems to be composed of rules, each rule has so many exceptions that it is of little use to know only the general rules.

What, then, is the difference between how those computer expert systems work and how human common sense proceeds? As I see it, this is a matter of the range and varieties of knowledge

they use. The expert system manages to work by using only a
few varieties of highly specialized knowledge about its subject
matter. Within each of those knowledge categories, the program
may "know" thousands of items, but they are all essentially of
the same type. However, the knowledge that a sensible person
must have to get around the ordinary world is not anything like
that: One has to know thousands of *different kinds of things.* So,
there is much more complexity in how things are represented
in our minds than in the computer programs. And this causes
a serious problem because working along the lines of our current
theories of artificial intelligence, it's difficult to discover good
ways of representing each kind of knowledge. The problem *can*
be relatively simple—even when there are huge amounts of data—
so long as we can use uniform ways to represent all of it. Computer programmers sometimes call these "data structures." The
trouble is, for common sense, we need many different kinds of
such structures, and no one yet knows any systematic ways to
link them together smoothly. And, without such links, machines
can't do ordinary, sensible reasoning. It's simply not enough to
program them with many separate facts: We also need good
ways to decide which facts to bring together, and good ways to
combine them.

How large are our human knowledge networks? No one really
knows, but I'd guess that it would take more than a million
linked-up bits of knowledge, but less than a billion of them, to
match the mind of any sage. (A billion seconds stretches thirty
years, and psychology has never found a way to make a person
learn something new each second for any prolonged period.) In
any case, understanding how to make machines that can build
such networks inside themselves seems to me the most exciting
research problem of our time. I believe that such problems are
difficult and complicated, but not impossible to solve.

It would be a mistake to try to study commonsense reasoning
without also studying the learning process. One reason for this
is practical: Even after we understand how to do it, it will still
be an enormous job to program into a machine all the knowledge
a reasonable person must have. To make our way around the
ordinary world, we each know how to recognize and use ten
thousand different kinds of things. Similarly, to make our way
through everyday conversations, we know several meanings for

each of a comparable number of words. To try to program that would be an awful job. In the end, it would be easier, and better, to program our machines to acquire such knowledge themselves: by watching what happens, by having conversations with knowledgeable people, by asking questions and making experiments, reading books, and doing all the other things that people do to educate themselves. Of course, it takes us humans time to learn. A person can become a good chess player or a mathematician in just a few years, but it takes even longer for an infant to become a competent adolescent. Still, once we discover an adequate set of principles for this process, there is no reason that intelligent computers could not learn much faster than we do.

There is another reason that machines must have the ability to learn in order to be truly intelligent. It is that knowledge is not a static thing. Whenever we solve any hard problem, some learning is involved—at least on a short time scale. True, we usually think of knowledge and memory as ways of storing thoughts away for *future* use. But in the course of solving any complicated problem, we also need ways to keep track of what has happened *recently,* so that we can change our strategies to get around the obstacles we have found. This requires the mind to keep records of what it has been doing; without such records, we'd keep going back to the beginning of each task and doing the same unsuccessful things over and over again. Furthermore, though we usually think of learning and memory as relatively passive, in the sense that they simply record what happens, consider what must happen after you solve a hard problem. In order to learn from such an experience, some part of your mind must have ways of deciding what to remember; it must know how to make guesses and form judgments about which features of the present situation are likely to apply again at later times.

How does human learning work? It would be wonderful if we could look to psychology and brain science for the answer. But really, these sciences still know very little about the mechanisms of learning and memory. As I mentioned, our very best methods are those that make use of comparative studies—the way the great Swiss psychologist Jean Piaget did in his research on how children think. As his research showed, children have different ways of thinking at various stages of development, and we can learn a lot, by comparing these different stages, about how we develop common sense. For example, certain skills seem never to appear by themselves, before certain others; this can lead us to suppose that later skills depend on the earlier ones in specific ways. Without such studies, by looking only at adults, we'd never see these internal dependencies. Piaget made many observations about the nature of the different kinds of thinking; some of these had never been noticed before, while others had always been taken for granted and assumed to be present from birth, rather than being developed throughout infancy and childhood. So those of us working in this field of machine learning should learn what psychology has to tell us, but I believe it has mostly been a one-way street going the other way. Scientists working on artificial intelligence are learning—even more rapidly than psychologists—about how knowledge is represented in intelligent systems, computers and brains.

Still, there is much to learn from biology, especially from the biology of the brain—for example, about how knowledge is stored in our brains. Are memory banks like freezer compartments in which time stands still, or do their contents slowly interact? How long do memories remain? Do some grow old and die? Do they get weak and fade away, or do they just get lost, never to be found again? We already know a good deal about this; for example, it appears that memories seldom become permanent unless their precursors are allowed to persist for about an hour or so. However, no one yet knows much about the nature of the machinery or processes that convert short-term memories to long-term ones. But we do know quite a lot about the speed of such processes. We know that the rate at which memories become fixed does not appear to vary very much among different people. Despite all those legends about prodigies, no good experiment has ever shown that any person can remember new facts—of any sort whatever—at rates faster than about one per second continuously for more than a few minutes at a time. Nevertheless, today there is no generally accepted theory of how memories are stored inside our brains.

What do we mean by "intelligence"? It never pays to try to make narrow definitions for things we don't yet understand very well, so let's just say we mean the ability to solve problems that people would say require intelligence. Still, one can ask, if intelligence is a single thing, could there be many clever ways for brains—or machines—to think? Of course, no one yet knows the answer to that, but it certainly would be surprising if there were only one single way. There are certainly many different ways to fly: Nature used mainly flapping wings, but humans have found other means as well—lighter-than-air balloons, propellers, and rockets, for example. Similarly, it could turn out that there are different ways for machines to solve problems. Indeed, some researchers in artificial intelligence hope to adopt some methods from logic and mathematics, in which everything else can be deduced from only a few basic principles. These theories look toward very neat, logical ways to make thinking machines.

Others in the field do not believe this can work. They argue that no such systems of logic can be found. Instead, they say that in order to make machines that can get good ideas, we'll have to give them the ability to use vague, approximate analogies. Such machines would not be designed around a very few, always applicable principles; instead, they would be engineered to accumulate large, and eventually huge, connections of observations and experiences. These analogy machines would then make themselves better and better able to guess which situations that have been encountered in the past are most similar to a new one and thus to deal with it effectively.

But, how could such a machine make decisions about which things are "similar"? As soon as one begins to think about this way of building intelligence—by using "thinking by analogy"— one sees that this is quite a problem in itself. Perhaps this is why so many researchers in AI have been entranced with the problem of making machines "see." For how we think depends very much on what we learn to see as similar.

Which colors do we think look most alike? Which forms and shapes, which smells and tastes, which timbres and pitches, which pains and aches, which feelings and sensations are similar? Such judgments have a huge effect at every stage of mental growth—since *what we learn depends on how we classify*. A child who classified fire only by the color of the flames might come to be afraid of everything colored orange. Then we'd complain the child had "generalized" too much. But if he classified each flame by tiny features never twice the same, that child would get too often burned and we'd complain he hadn't generalized enough. This problem of similarity is so important that, as the reader will discover, questions about visual recognition and classification have turned into a whole field of research. In everyday life, we take for granted the ability to recognize things and never realize how complicated seeing really is.

Isn't vision just a peripheral accessory? After all, blind people can be just as smart as sighted ones! Nevertheless, there are good reasons to focus on vision in artificial intelligence. The obvious one is that, as far as robots are concerned, vision will be extremely useful. This isn't just because the better a robot can see, the easier it will be for it to do things. Vision will also be useful because it will enable us to show—that is, to teach—our robots what to do. This is important; everyone who has had anything to do with computers knows how hard it is to "tell" them things. How much easier this will be, in the future, once we can demonstrate by example. (When one comes to think of it, the same is true with ourselves.)

But practical aspects aside, vision research has actually turned out to be important in more basic research on theories of intelligence. Indeed, perhaps our general ability to think so well has evolved from our ancestors' ability to see so well.

This raises the possibility of yet another approach to how to make intelligent machines: to copy human psychology. This is the approach I prefer; the only trouble is, as I mentioned, we still don't know enough about psychology. Nevertheless, since we're the only creature around that does the kinds of intelligent things we want our computers to do, we may as well use ourselves as examples.

Is there any reason to think that the first way evolution found to make intelligent machines—the human way—is a particularly efficient one? Well, yes . . . and no. Yes, because evolution generally tends to try the simpler combinations first. And no, because much of evolution is a matter of chance, it doesn't *always* try the simplest or best combinations first. Furthermore, when evolution finds a combination that works, it tends to stick with it, because whenever a species tries a larger change, things almost always get worse before they get better. This means it is nearly impossible for an evolving species to make any really basic changes in its older structures; the competition to survive is usually too fierce to leave enough of the temporarily less-efficient survivors to ensure that the change will be passed on.

This probably means, at least on this particular planet, that there is essentially no chance that other animal species will evolve any very different or more advanced forms of intelligence in the course of natural evolution. The only likely way that could happen would be as a result of our own careful supervision, either in the form of controlled genetic modifications or, more likely, as planned designs for intelligence in machines.

One thing few outsiders are told is just how long it takes to solve these kinds of problems. The new sciences of robotics and artificial intelligence are very young, barely twenty-five years old. A mere handful of people were working on these profound problems in 1960. Today there are perhaps a few hundred, but that is still very few compared to fields like physics or molecular biology, which have literally tens of thousands of research workers. And, as in every other field of research, most of these people are working on relatively immediate, practical aspects. Only a small fraction of them work on the hardest, most fundamental, long-range problems. Accordingly, this means that in the new science of artificial intelligence only a hundred or so people are involved in what one would call genuinely basic research. This has produced an interesting situation. There are many hundreds of important theoretical problems, but far fewer scientists working on them. Therefore, at any given moment of time, there are many important problems that no one is working on.

Why are there so few individuals doing basic research? It's not entirely a matter of having the imagination to do that sort of work. It also requires a certain special kind of temperament. Few people have the disposition to take on problems they aren't sure can ever be solved—there's so little certainty of any payoff or reward. You have to have a certain arrogance and independence to try—perhaps for years—to do something entirely new. In that sense, the most ambitious scientists are often a lot like gamblers.

Another little-known fact is that it is not always the well-known scientists who do the most important work. Sometimes, of course, they do, but at other times the senior person's name appears as if by magic on the final publication, which is all the public will ever see. However, in embryonic fields like robotics, where ideas come and grow and fade away so rapidly, much of the most important work is done by relatively young students.

I would say that more than half of the basic discoveries in AI have been made by students before they completed their Ph.D. theses. Afterward, quite a few of those brilliant young people abandon research and instead work on practical applications. Despite all the promise of and publicity about this field, there are few opportunities after one leaves the university to continue in basic research. On the whole, industry is well intentioned about the future of robotics, but it simply doesn't understand the problems.

Here is a typical scenario of the time scale of basic research in artificial intelligence, one based on my own experience. In the late 1950s, I had a certain idea about how computers could be used to describe things. It took me two or three years to clarify this idea, and it was published in 1961. (I was lucky—as it happened, one of the professional journals was rushing out a special issue about computers. Normally, it would have taken me four or five years to get published.) In the meantime, I became a professor and started to try to convince my students to work on such ideas. In 1964, one of my students, Thomas Evans, finished a beautiful piece of research, which was partly based on my work but included many more ideas that were entirely his. What Evans did was to show how my description idea could be used to make a machine discover certain kinds of analogies between different things. Thus seven years had passed between a first conception and a demonstration (by someone else) of how it could be used. Why did it take so long? Because it needed a new way of thinking.

What happened next? Nothing much. I explained to every new student the importance of what Evans had done, but each of those brilliant people had one or another different way of thinking. Some couldn't see how to use what had already been done. Others did not see that it was especially important. Each of them became attached to some other good ideas and did outstanding work on them. After all, what else could one expect: There were perhaps a hundred good ideas around, but only a couple of dozen researchers! Today there are probably a thousand good ideas around, and still the number of good researchers is less than that.

In any case, a few years later, another student clearly saw the next step to take in this area, and in 1970, after several years of work, the head of the AI lab at MIT, Patrick Winston, finished a wonderful thesis; in it he showed how to combine the earlier ideas—and many important new ones, too—to make a machine that could *learn to find analogies*, at least to a limited extent. I found this tremendously exciting, because it was clear to Winston and to me that this could unlock many other mysteries as well. Then, for the next few years, both of us taught this theory to all the new students. And again the same thing happened; for a long time no one was entranced by this particular set of ideas. Why not? For one thing, there were several other equally wonderful ideas competing with them, and some of them appeared to be both easier to solve and more profitable. For another, Winston and I were engaged with other matters. Finally, even when a new person decides to take on such a project, it takes two or three years to gain the experience to handle it, another two or three to carry out the project, and then yet another year or two to understand what was done clearly enough to write it down. Only then can it be explained precisely, in a way that will stimulate someone to start the cycle over again. Most people simply cannot tolerate that pace: It seems too slow. Yet on the scale of history, it's very, very fast indeed. From Newton to Einstein was only two and a half centuries. The great developments of modern physical science all took place in less time than it took to build a single great cathedral in the Middle Ages.

It is funny how impatient people get. For example, about twenty years ago one of AI's great pioneers, Herbert Simon of Carnegie-Mellon University, predicted that in another ten years computers might play chess so well that some machine could be world champion. When the decade ended, some angry critics shouted that he had been wrong and that no machine would ever be that good at chess—indeed at the end of that first decade, the best chess machines could scarcely yet play decent, honest, amateurish chess. Even at the end of the second decade, those critics still complained, for now the best chess machine could merely play at near master level and could beat only about 999 out of 1,000 good chess players. What a disappointment! Now it is perhaps time for a new prediction: It may be still another decade before a chess machine gains the status of international grandmaster, and two decades to the world championship. It seems to me the principle is the same, in any case.

But it might be more illuminating to ask why Simon's prediction came out wrong. I think the answer is just so simple that it was overlooked: The error was not technical, but economic. As it happened, only three or four people actually pursued research on chess strategies during that period, and there was no inducement to get others to join them because there was no obvious profit in it. And because there's no way to prove me wrong, I'll add this little bet: If Simon himself had spent ten years on it, I'm sure his prediction would have been right on the nose.

Now let's consider the time factor in relation to a hard problem that I've already mentioned and that some of us have been working on for the past twenty-five years: how to make computers have common sense. In the near future I will publish what I've learned about this, in a book called *The Society of Mind.* I believe it will contain some good ideas, but even if that's true, it will take another quarter of a century to separate the good ones from the bad ones. If it takes a quarter-century to test an idea—or to find out that it wasn't so good after all—then how long may it take to make machines truly intelligent? If we need just twenty such ideas and each one has to be developed on the basis of the last, then the process could take 500 years. And though so far as history's concerned, that's very little time, it may seem too long for mortal people to be bothered with. But what if some of those ideas are independent of one another—so that they could be studied, each one by a different research group, all at the same time? If that turns out to be the case, then we may already be very close to making highly intelligent machines. The future no one knows. It could be a long way off—or it could be just around the corner!

It is also not out of place to note that the artificial intelligence community has been responsible for a surprising proportion of the advances in computer science. Not only did the time-sharing concept come from them, but so did many of the advances in word processing and office automation, the new LISP machines that are used in advanced development centers today, many of the computer-aided design and graphic systems now becoming popular—and soon a great new wave of multi-parallel and so-called fifth-generation designs. (True, no one knows what the first four generations were, exactly, but who cares? It has a nice sound.)

Today we in AI could use more support, but I see no real reason to complain that industry does not contribute enough to basic research. Often when a company decides to make such a move, it may be too far ahead of time. When the AMF corporation first produced commercial assembly-line robots, it was unable to market them successfully, because few industrial production managers could figure out how to use them. In my view, commercial robots have not really come very far. There still are no industrial robots that can begin to approach the sensitive, refined motions that every person's hands make in such mundane tasks as tying a shoelace. Basic research on gaining more dexterity is proceeding only very slowly, here and there, because such serious projects are so very few.

One of the problems with seeking support for basic research is that it is hard to predict just who will be the beneficiaries of any particular area of research. I want to explore two examples of fields still in their early infancy. First, suppose that we could develop a really powerful vision system—that is, a machine that could really "see" in the sense that people see. By this, I mean the ability to look around a room and recognize the presence of and relations among all the ordinary objects there: to be able to say, "There's a desk, with books and papers and envelopes and paperclips; there's a person sitting in a chair; that's a window and a real tree—and *that's* a picture of a tree on the wall." If we had such a machine, what could we use it for? That's simply an impossible question! One would do better to ask, "What could we *not* use it for?" One can scarcely think of any place where such machines could not be used.

A second field that will surely change the world is the one I call telepresence. At some point in the future someone would go to work by slipping on a comfortable jacket lined with a myriad of sensors and musclelike motors. Each motion of his arm and fingers would then be reproduced at another place by mobile, mechanical hands. Light, dexterous, and strong, those remote mechanical hands have their own sensors, which will transmit what's happening back to the worker so that he will seem to feel whatever the remote hands may touch. The same will be done for the motions of the head and eyes, so that the operator will seem to see and sense what's happening in the other workplace. Once we can do such things, it will be another simple step to give those remote presences different strengths and scale of size. These remote bodies can have the brute capacity of a giant or the delicacy of a surgeon. And, using these information channels, an operator could be anyplace—in another room, another city, another country, even out on a space station orbiting the Earth.

What will telepresence do? Among its tasks would be included:

Safe and efficient nuclear-power generation, waste disposal, and engineering, even under the sea. We all saw the absurd inflexibility of today's remote-control systems in the Three Mile Island break-down. Surely telepresence could have averted much of the danger, the damage, and the expense of repairing the injured reactor.

Advances in fabrication, assembly, inspection, and maintenance. With telepresences one can as easily work from a thousand miles away as by being right there. We could invent new kinds of "work clubs" in which several people could combine their part-time energies.

Elimination of most chemical and physical health hazards, as well as the creation of new medical and surgical techniques. If we miniaturize mechanical hands, then microsurgery would be much easier, and surgeons could repair more tiny tendons, ducts and other structures now beyond our clumsy reach—especially the inaccessible areas of the brain. And as for industries that are inherently hazardous, at some time in the future, if a tunnel or mine should collapse, we can say unfeelingly, "Too bad. We've lost a dozen robots."

Construction and operation of low-cost space stations. The United States is now committed to build a permanent space station in the early 1990s. How much more useful and effective it would be if we could design it from the start to exploit telepresence technology.

Naturally it will be a while before we see such robotic feats. Today our robots and our primitive AIs are still much more like wind-up toys than like our own great human minds. However, we're seeing signs of things that real robots ought to be able to do, like sensing the sounds of certain words and acting on those phrases in accord with real—although limited—knowledge of what they signify.

As progress continues, we'll reap the fruits of our research and start to see machines that display more genuine signs of having minds. We'll start to give them learning skills to organize their little minds, so that they can learn from us, and from each other, as we do. We'll show them how to make copies of themselves. Most of them won't even have to learn such things, because they will be manufactured already knowing them. We'll give them limbs even more dexterous than our own and new kinds of senses that will seem to us uncannily observant. Gradually, they'll begin to slip across that edgeless line of doing only what we programmed them to do and begin to move themselves into the zone of things that *we* are programmed to do. Then, of course, mere telepresence will be seen as having been only a passing stage, until the brainy AI machines became smart enough to do the jobs themselves.

What will happen when we have to face those options in which intelligent machines begin to do the things we ourselves *like* to do? Economists and sociologists are helpless in the face of such questions, because their sciences assume that the entities they're dealing with—that is, people—will remain essentially the same. The only thinkers in our society who, in my view, have really tried to discuss such drastic technological changes are the writers of science fiction. Unless we decide to ban AI, to try to keep the world the way it is today, we will all someday have to face those issues that so far only science-fiction writers have raised. What kinds of minds and personalities should we dispense to our machines? What kinds of rights and privileges shall we give them or withhold from them, and what roles shall we allow them in the societies which, up to now, we alone have ruled?

This book is for those who have the courage and disposition to consider such questions. For the most part it will not answer the questions but only make them seem harder than before. But that's the way the problems we have to face ourselves should be—now that we are on the verge of leaving all the easy questions to our machines.

Man Makes Man
T. A. Heppenheimer

The art and occult science of creating a manlike creature has passed from the alchemists and intuitively gifted individuals of the Middle Ages to the mathematicians, physicists, and engineers in the twentieth century. There is an almost mythic bond between the two, the continuing fascination with robotics, the challenge of man making man.

There is a story told about Friar Roger Bacon, a brilliant English clergyman-scientist-philosopher of the thirteenth century, and his colleague Friar Bungey. The tale said they wanted to surround England with a wall of brass for protection against invaders. To do this, Bacon proposed first building a brass head that would explain how to build the wall. This done, he summoned a spirit to give the head the power of speech. The spirit warned Bacon that if he and Bungey did not hear the words of the head when it spoke, all would be lost.

For three weeks the friars sat watching the head, but it was mute. Finally, they decided to get some sleep and engaged a guard to watch, instructing him to wake them should the head speak. No sooner had they gone to sleep than the head spoke for the first time: "Time is." To the guard this seemed too trivial to justify waking the friars. Half an hour later the head spoke again, saying: "Time was." This also seemed too trivial to take seriously. After another half-hour the head declared: "Time is past," and collapsed.

The tale is apocryphal, but Roger Bacon was a particularly apt choice for a story like this. He was a bold thinker, an insightful scientist, a genuine futurist, a Renaissance man before there was a Renaissance, and a master of most of the learning of his day. It was Bacon who wrote that it would be possible to reach India by sailing westward from Spain. It was he who declared an arrangement of lenses would allow close study of the moon and stars. It was also Bacon who predicted science would someday give us machines for flying, machines that would let us move at incredible speeds without animals, machines that would let us navigate over water more swiftly than ever conceived, and machines that could take us to the bottoms of seas and rivers without danger. Seven centuries later there is a generation of scientists—the intellectual progeny of Friar Bacon—keeping watch in their robotics and artificial intelligence laboratories, waiting for the "Time is" call of their creations. They are determined not to miss it.

The art and occult science of creating a manlike creature has passed from the alchemists and intuitively gifted individuals of the Middle Ages to the mathematicians, physicists, and engineers in the twentieth century. There is an almost mythic bond between the two, the continuing fascination with robotics, the challenge of man making man.

Legends and lore about man-made creatures are behind much of the ancient fascination with the subject of making a manlike object. There is a rich tradition of "talking head" tales like the one told about Roger Bacon. Similar talking heads were supposedly built or owned by Albertus Magnus, the teacher of Thomas Aquinas; by Bishop Grosseteste, a philosopher and contemporary of Roger Bacon; and by Pope Sylvester II, among others. With advances in the skill of the alchemist, there were those in the sixteenth century who declared it was possible to create not merely a brass head, but an entire living human being, called the homunculus. The medieval physician Paracelsus, one of the leading doctors of his day, gave the recipe:

> Let the semen of a man putrefy by itself in a hermetically sealed glass with the highest putrefaction of horse manure for forty days, or until it begins at last to live, move, and be agitated, which can easily be seen. After this time it will be in some degree like a human being, but nevertheless, transparent and without body. If now, after this, it be every day nourished and fed cautiously and pru-

dently with the arcanum of human blood, and kept for forty days in the perpetual and equal heat of horse manure, it becomes thenceforth a true and living infant, having all the members of a child that is born from a woman, but much smaller. This we call a homunculus; and it should afterwards be educated with the greatest care and zeal, until it grows and begins to display intelligence.

All this was based on alchemical lore. The manure represented mother Earth; its putrefied state followed alchemical tradition. The recipe reflected Aristotle's view that an embryo is formed from semen combined with menstrual blood, the flow of which ceases during pregnancy. The hermetically sealed jar took its name from Hermes Trismegistus, "the thrice-great Hermes." The forty days reflected the law of the Church, which declared that a fetus would quicken in the womb forty days after conception.

We can appreciate Paracelsus's influence by considering that his true name was Theophrastus Bombastus von Hohenheim, and that his family name gave our language the word *bombast*. Nevertheless, his recipe contributed to the legend of Faust, who sold his soul to the devil for the pleasures of love. In Goethe's great play of that name, Wagner, Faust's assistant, creates a "graceful, dazzling little elf" by alchemy. But this was little more than a detail in Goethe's narrative, a way of displaying Faust's magical powers.

In another set of legends, the man-made man again took center stage. This was the lore of the golem, which arose in the second half of the sixteenth century, several decades after Paracelsus. The word *golem* is Talmudic; it refers to anything incomplete or unformed, such as an embryo or the shapeless mass of dust from which Yahweh created Adam. Around 1550, Elijah of Chelm was said to have created an artificial man, called a golem, with the aid of the ineffable Name of God, the four Hebrew letters corresponding to YHWH. It is said to have become a monster threatening the world, until the sacred name was removed. Elijah's grandson, in the fashion of Talmudic scholars, debated the question of whether this golem could count as one of the ten men in the *minyan*, or quorum, needed to conduct a religious service.

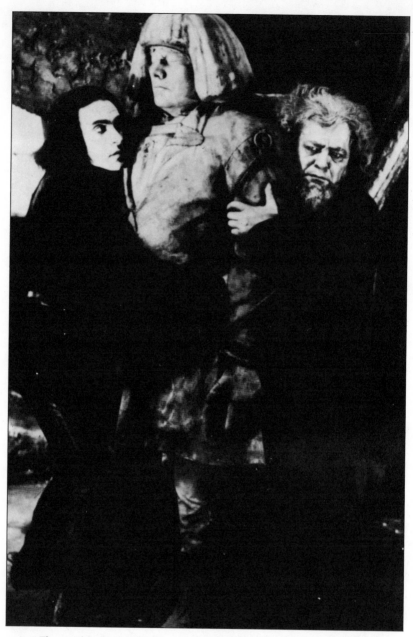

The mythic roots of robots are found in the story of the golem,
the man-made creature portrayed here in a 1920 German film.

Thirty years later came another legend of a golem. This was supposed to have been the work of the Chief Rabbi of Prague, Judah ben Loew, known rather irreverently as "the High Rabbi Loew." Rabbi Loew was a historical figure, a friend of the astronomers Tycho Brahe and Johannes Kepler. In real life he was a sober theologian, not a man to meddle with magic, but the legend about him was rather different. To protect his people against the pogroms, the tale goes, Loew and two assistants went in the dead of night to the River Moldau, and from the clay of the riverbank they fashioned a human figure. One assistant circled the figure seven times from left to right. Loew pronounced an incantation, and the golem began to shine like fire. The other assistant then began his own incantations while circling seven times from right to left. The fire went out, hair grew on the figure's head, and nails developed on its fingers. Now it was Loew's turn to circle the figure seven times, as the three of them chanted words from Genesis. When Loew implanted the Holy Name upon its forehead, the golem opened its eyes and came to life.

It was incapable of speech, but had superhuman strength. This made it useful in defending the Jews of Prague against the Gentiles. It was also Loew's servant and worked as a janitor within the temple, though it was allowed to rest on the Sabbath. Only Loew could control it, but eventually the golem could not be controlled at all. It ran amok, attacking its creator. Its career of destruction ended only when Loew tricked it into kneeling before him, whereupon the rabbi plucked the sacred name from its forehead. Magically, the golem was once again reduced to clay.

These "artifacts"—the brass head, homunculus, and golem—all belong to what might be called the anthropology of robotics. There is no hint of any technical means adequate for their creation; rather, this was to be accomplished by magic. These legends are important, nevertheless, for they separate cleanly the idea of the man-made man from its technical background, allowing us to focus on its psychology. The predominant idea is that, in the hands of a savant, anything in the form of a man will have the power of a man. What is interesting is that within this anthropology, we can discern an evolution. Bacon's head was no more than a sculpture with the power of speech. Loew's golem was to work as a servant, but turned on its master. With a small admixture from the physical sciences, this golem, in the hands of Mary Shelley, would become Frankenstein's monster.

Concurrent with this robot paleontology runs a prehistory of the robot as machine, a sequence of inventions both real and mythical. Within the prehistory, the dominant theme is the automaton, or mechanical man. Its technical background can be traced to the ancient Greeks. The prototype of a robot's grippers was the mechanical claw of the catapults of antiquity, which substituted for an archer's fingers, engaging a massive bowstring so that it could be pulled back and then released with a trigger. These catapults were both sophisticated and deadly. When the Spartan general Archidamus saw one being fired, he declared, "Oh, Hercules, human martial valor is of no use anymore!"

The gears and cams of the earliest true mechanical technology first developed in the Middle Ages, with the windmills and water wheels used to grind grain. Pneumatic bellows, important in many early automata, was a feature of the first pipe organs. In Winchester Cathedral, over a century before the Norman Conquest, an organ was installed having twenty-six bellows and four hundred pipes. Its keyboard featured keys so enormous that the organist had to strike them with his fists, protected by thickly padded gloves.

The first mechanical men were the moving figures on the clock towers of the late Middle Ages. Above the Piazza San Marco in Venice stand two large figures with hammers that strike a bell on the hour. More impressive was the clock of Strasbourg, built in 1574. It featured a cock or rooster of cast iron, a reminder of St. Peter's denial of Jesus. It appeared at noon in the company of twelve figures representing the Apostles, opened its beak, stretched out its tongue, flapped its wings, spread out its feathers, raised its head and crowed three times, the crowing being produced by bellows and a reed. In use until 1789, it inspired Hobbes, Descartes, and Boyle, all of whom saw it as a specimen of what might someday be accomplished by machinery.

To some degree, the use of machinery was merely magic in a different guise, a way of rendering plausible what in fact was not yet possible. Thus, the fifteenth-century astronomer Regiomontanus was rumored to have built an artificial eagle, which flew to greet Emperor Maximilian as he approached Nuremberg in 1470, then returned to perch atop a city gate. Leonardo da

The Strasbourg cock, a cast-iron rooster, was one of the wonders
of its day. Symbolizing St. Peter's denial of Jesus,
it appeared at noon each day and sounded the hour by opening its beak,
spreading its feathers, flapping its wings, and crowing.

Man's fascination with manlike machines is ancient.
This is how seventeenth-century designers rendered the animated statues
of one ancient inventor, Hero of Alexandria, who lived in the fourth century B.C.
The figure of Hercules raised his club to strike the dragon on the head.
When hit, the monster spit water in his face.

Vinci supposedly built an artificial lion to honor the King of France, Louis XII, around 1500. As the king entered Milan, the lion advanced toward him, then opened its chest with a claw, displaying the fleur-de-lis, the French coat of arms. About 1640, Descartes is said to have constructed an automaton that he called "ma fille Francine" and kept in a case. During a sea voyage, the ship's captain allegedly saw it move like a human being, declared he had witnessed the work of the devil, and threw it overboard.

Many of the early automata were the works of artisans at the courts of kings, seeking to provide their monarchs with unusual toys or diversions. Around 1540, Gianello della Torre of Cremona fashioned a figure of a young girl playing the lute, to alleviate the boredom of Emperor Charles V. She could walk in either a straight line or a circle, while plucking the strings and turning her head from side to side. Early in the seventeenth century, two French engineers, Isaac and Salomon de Caus, built a number of ornamental fountains featuring moving figures. In one of their most famous works, a bird appeared on a branch and sang with the sounds produced by air being blown across a water-filled pot; but when a mechanical owl appeared on a nearby rock, the bird would disappear in apparent terror.

Meanwhile, the art of clockmaking was advancing rapidly. What encouraged it was the search for a clock of sufficient accuracy, which would be invaluable for determining longitude at sea. In 1759 watchmaker John Harrison presented the British Admiralty with such a clock, called a chronometer, the result of nearly fifty years of development. It was granted a sea trial, the first destination being Madeira. Nine days out from Plymouth, the ship's longitude, by dead reckoning, was 13°50'; but according to Harrison and his instrument, it was 15°19'. The captain favored his dead reckoning calculations, but since he was under orders from the Admiralty he followed Harrison's directions. Harrison declared that if Madeira was properly marked on the navigation chart they would sight it next day. The captain bet Harrison five to one that he was wrong, but he held the course. Next morning, at 6 A.M., the lookout spotted Porto Santo, the northeastern island of the Madeira group, dead ahead.

In a later test before King George III, Harrison showed that his chronometer was accurate to within 4½ seconds over a period of 10 weeks. This phenomenal precision was achieved within a mechanism measuring only 5 inches across.

Several of Harrison's contemporaries applied this technology to create artificial humans and animals of surpassing realism. One of the most famous was the duck of Jacques de Vaucanson, first displayed in 1738. It could quack, splash around in the water, eat, drink, subject its food to a chemical change intended to represent digestion, and then excrete this material through an anus. In the words of a contemporary account: "After each of the duck's performances there was an interval of a quarter of an hour to replace the food. A singer announced the duck. As soon as the audience saw it climbing on the stage, everybody cried, 'Quack, quack, quack!' Greatest amazement was caused when it drank three glasses of wine."

The duck would stretch its neck to pluck grain from its keeper's hand, and it imitated the gestures a duck makes when swallowing quickly. Its wings alone each contained more than four hundred articulated pieces because, as Vaucanson put it, it was his intent to duplicate a real duck's wings "bone by bone." His intention was not only to demonstrate his machine, but to exhibit its workings; thus, he could open the body and expose the duck's inner mechanism to view. Nevertheless, some ladies preferred to see it remain decently covered.

Vaucanson also built two androids, mechanisms in human form. Both were musicians; one played the flute, the other, the drums. People found it hard to believe that the flutist was actually playing, rather than producing the musical sounds from a hidden set of instruments. But the flutist's breath did come directly from its mouth by means of a bellows; a mechanism regulated its lip movements, and the flute was a standard one. By making finger motions over the holes of the instrument, the android could play a repertoire of twelve tunes. In the prehistory of robots, this was a landmark. A relatively small number of people can play the flute well; Vaucanson's flutist can be described as the first mechanical device to excel over most people at the performance of a learned skill.

With machines being built with great cleverness to imitate a human, it was just a matter of time before someone arranged that a human would mimic a machine. This was the work of Baron Wolfgang von Kempelen, an eighteenth-century inventor.

One version of the story has it that he was at the court in Vienna and was asked by the Empress Maria Theresa to tell her about a then-current rage, hypnotism. Replying that he had something better to offer, he produced what appeared to be a chess-playing machine. Another version adds that he built this to help a legless and accomplished chess master to escape from Russia. Whatever the rumors about it, Kempelen's machine was real enough. It featured a mannequin in the form of a Turk, with turban and handlebar mustache, seated behind a wooden cabinet that supposedly held the mechanism. When demonstrating it, Kempelen would ostentatiously open the doors of the cabinet, one after another, to show that no one was hidden inside. Actually, by moving around within the cabinet, the Polish chess master (if he did exist) could easily have kept out of sight.

With great aplomb, Kempelen would wind up the machine's clockwork and invite a member of the audience to play. To the whirring sound of machinery, the Turkish mannequin would pick up the chess pieces and make its moves. If its opponent made an illegal move, the Turk would shake its head and replace the piece. If it threatened the opponent's queen, it shook its head twice; if it put its opponent in check, it gave three shakes. The Turk won games all over Europe. A number of authors wrote about this wonderful chess player, including Edgar Allan Poe; most logically concluded that there was a man inside. Still, this likely fraud was another milestone in the early development of robotics. For the first time, an inventor had succeeded in blurring the distinction between man and machine.

Kempelen produced his chess player in 1769. Between 1770 and 1773, a father-and-son pair of collaborators, Pierre and Henri-Louis Jaquet-Droz, demonstrated three amazing human figures known as the Scribe, the Draftsman, and the Musician. All three were operated by clockwork and featured intricate arrays of cams. The Scribe and Draftsman were in the shape of young boys, elegantly dressed. The Scribe could dip a quill pen in an inkwell and write a text of up to forty letters. His hand could move in three directions; each direction was controlled by a cam, and each turn of a stack of cams would form one letter. By setting levers on a disk used for control, the Scribe could be made to write any desired text. His brother, the Draftsman, could execute drawings of Louis XV and similar figures; a later

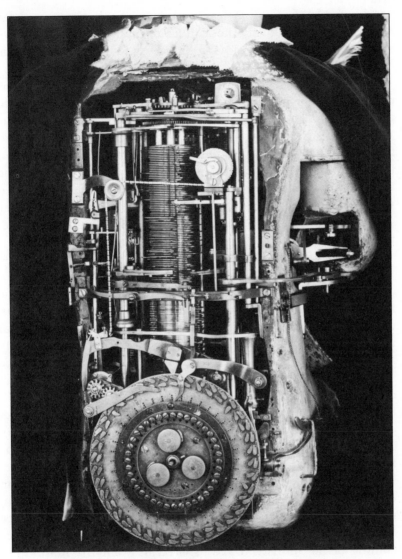

The key to the astounding versatility of this Scribe automaton
(it could write text up to forty letters) is the elaborate system of
cams and levers that moved it through its preprogrammed motions.

version could draw a battleship under full sail, showing three rows of gunports. Both of these androids maintained an attentive attitude while at work, moving their eyes to follow the tracings of their hands.

Another Jaquet-Droz android, the Musician, resembled a girl of sixteen, wearing a powdered wig and a dress appropriate to the court of Vienna. She played the organ—*really* played it, as Vaucanson's flute player had done. She moved her arms and fingers to push the keys in the proper order. Her chest rose and fell to simulate breathing; she moved her body and head in time to the music; her eyes glanced about in a natural way. Even when she paused in her playing, these movements made her seem alive. At the end of her performance, she would do a graceful little bow.

These automata still exist, and remain in working order. The Jaquet-Droz devices are in the Musée d'Art et d'Histoire in Neuchâtel, Switzerland. The Draftsman with its battleship is at the Franklin Institute in Philadelphia. In these inventions we can discern many of the features of today's programmable industrial robots. In particular, they executed accurately and repetitively a detailed sequence of motions, controlled by a program (the stack of cams). We could say that the industrial robot is a direct descendant of the Scribe or the Musician, with only three differences: the adoption of a functional form rather than the human one, its use of hydraulics and other power sources instead of springs and clockwork, and the use of programming methods more sophisticated than cams.

By 1800 the Industrial Revolution was in full swing. The steam engine would soon power trains and steamships, and the European world was poised for the outstanding advances in science and engineering that would set off the age. Yet for robotics, the coming of the industrial age ushered in a lengthy hiatus. For about a century and a half, roughly from 1790 to 1940, there was virtually no significant advance in this art.

Artists, from ancient times to the Renaissance, had mastered the technique of simulating human form—the art of sculpture. The medieval and Renaissance builders of automata had created a new art—reproducing human motions—and developed its techniques to a high degree. But to go further, to advance beyond blind preprogramming, it would be necessary to give a machine the power to make decisions. That meant it would have to store and handle information, and to loop and branch among possible alternative paths within a program. This would not become possible until World War II; even then, a proper understanding of automated decision-making would await the seminal insights of a brilliant mathematician, John von Neumann.

The story of Charles Babbage illustrates why this new development took so long. Babbage, a nineteenth-century British inventor, had a clear idea for a general-purpose digital computer featuring a stored program. He called this concept the Analytical Engine, and he knew that it would be able to solve a wide range of problems in mathematics and analysis; it would even be able to play chess. However, its technical challenge defeated him. He sought to build it using the clockwork technology of the day, but his machinists could not make parts with the needed precision. His demands for exactitude in the engine's ratchets and gears, however, helped to bring the machine-tool industry into existence.

It's not that nineteenth-century engineers didn't have the means to achieve the automatic storage and handling of information. As early as 1788, James Watt devised a flyball governor, featuring two whirling balls able to swing outward by centrifugal force. It was linked to a steam engine, and the outward swing of the flyballs measured the engine's speed; but more than that, by another linkage this outward swing controlled a valve that maintained a preset speed. It was, in short, a simple feedback control mechanism—the world's first. In 1868, James Clerk Maxwell (the discoverer of Maxwell's equations in electromagnetism) saw that there was something new in such a device. His paper, "On Governors," was the first systematic study of feedback control, which would be essential in devising the robots of the twentieth century.

Then, in 1886, there was Herman Hollerith's tabulating machine. Like Harrison's chronometer, its invention was driven by necessity. The U.S. Census Bureau needed a more efficient method of tabulation; it had taken seven years to turn the raw data of the 1880 census into finished sets of tables. The Constitution requires a census every ten years, and it was clear that, unless machinery came to the rescue, a new census would be due before the last one was tabulated. Hollerith, a statistician, devised a rapid method to handle the data. He worked out a system whereby it would be coded and punched on cards; such cards had been proposed by Babbage and were adapted from the automatic loom invented by Joseph Jacquard in 1801. The cards were carried, one by one, over a pool of mercury. A set of wires dropped onto each card; the one that found a hole would drop through to the mercury, making an electrical contact to a counter, causing it to advance by one unit. The recorded data then could be read right off the counters. Hollerith had solved the census problem: The 1890 census was digested to finished tables in two and a half years. To market his tabulator, he founded a company that later became International Business Machines.

And there was the automatic chess-playing machine of Leonardo Torres y Quevedo, in 1912. It would play an end game of king and rook against king. For this game, an explicit set of rules can be given and checkmate accomplished in a limited number of moves, regardless of the initial positions of the pieces. This probably stands as the first machine not only capable of handling information, but also of using it to make decisions. As Quevedo wrote in 1915, "The limits within which thought is really necessary need to be better defined . . . the automaton can do many things that are popularly classed with thought."

In this meager technological progress in robotics, the main contributions were literary: Mary Shelley's *Frankenstein* in 1817 and Karel Čapek's 1921 play *R.U.R.* Shelley had seen the Jaquet-Droz automata in Switzerland and was familiar with the lore of the golem. Her story amounted to an update of the golem legend; her monster was brought to life, however, not by alchemy, but by the science of her day. Her leading character, Victor Frankenstein, was an idealist, a scientist bent on pursuing knowledge for the sake of bringing new and useful discoveries into the world. However, he did not reckon with the possibility that something might go wrong.

Frankenstein makes a creature eight feet tall and brings him to life on a rainy November night with the aid of electricity. (The use of electric currents to make frogs' legs twitch was then a recent discovery.) The young scientist expects all the love and kindness that a father might get from a son; but when he sees the repulsiveness of his creation, with its watery eyes and ill-fitting skin, he flees in horror. The creature, left on its own and bereft of Frankenstein's fatherly love, goes berserk and murders Frankenstein's little brother. Frankenstein pursues it, they meet and talk. The monster demands that Frankenstein create a companion for it. He agrees and sets to work, but then has second thoughts; what might these two monsters do together? With Frankenstein's promise broken, the monster again runs amok, murdering the scientist's best friend and causing the death of his father. Even so, Frankenstein is compelled to hold to his quest for knowledge, in hope that it will bring a better world. At the end, the monster presents itself before its creator and begs forgiveness.

If *Frankenstein* was a novel of ambivalence, *R.U.R.* was a cry of despair, an echo of the agony of World War I, which had brought all the evils Mary Shelley could ever have dreamed of. That war had started conventionally enough, with columns of German foot-soldiers marching into France, accompanied by horse-cavalry and artillery. Very soon, however, it degenerated into a bloodbath, a massacre, made possible by such inventions as the machine gun, the submarine, and poison gas. Such technology not only intensified the carnage but prolonged the conflict, which tore Europe apart more thoroughly than any conflict since the Thirty Years War (three centuries earlier). In its wake, *R.U.R.* gave us the word *robot*—a name, an image, and a reputation to live down.

The name *robot* came from Čapek's Czech language, from *robota*, meaning labor or work. Other Slavic tongues have similar words (the Russian *rabota*, for example); they can be traced to the Indo-European and are cognate to such words as the German *arbeit*. In Czech, however, *robot* carried connotations of slavery or forced labor.

The image of a mechanical robot, at least, was less sinister. It could be traced to a 1624 Italian work by Giovanni Battista Braccelli, *Bizzarie di varie figure*, whose drawings of acrobats and dancers presented images of blocky, angular, metallic men built from machine parts. Similarly, there was the "steam man" of a Canadian engineer, George Moore, in 1893. It amounted to a walking locomotive; it featured a gas-fired boiler in its belly and puffed smoke from a smokestack in the shape of a cigar.

R.U.R. is a play about Rossum's Universal Robots, artificial men and women whose design has been simplified to permit

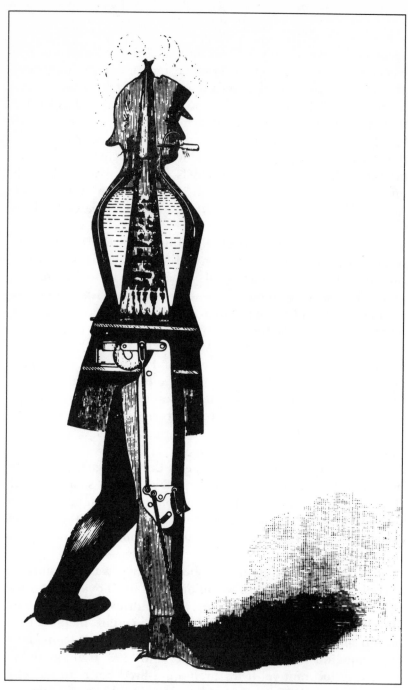

This automated anachronism—a knight in armor smoking a cigar—
was a steam-powered walking machine. Canadian inventor
George Moore installed a gas-fired boiler in its belly to make it go.

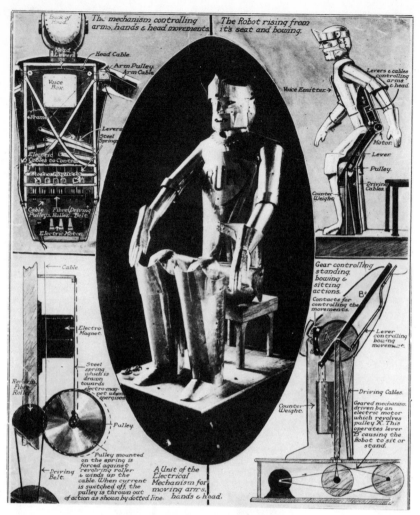

The mechanism controlling arms, hands & head movements.

The Robot rising from it's seat and bowing.

Back of Head.

Head Levers.

Head Cable.

Arm Pulley. Arm Cable.

Voice Box.

Frame.

Levers.

Steel Spring.

Electrical Cables to Control

Electro Magnets

Cable Pulleys. Fibre Driving Roller. Belt.

Electric Motor.

Voice Emitter.→

Levers & cables controlling arms & head.

Motor.

Lever.

Pulley.

Driving Cables.

Counter Weight.

←Cable.

←Electro-Magnet.

Steel spring which is drawn towards electro-magnet when energised.

Revolving Fibre Roller.

←Pulley.

Pulley mounted on the spring is forced against revolving roller & winds up the cable. When current is switched off, the pulley is thrown out of action as shown by dotted line.

←Driving Belt.

Gear controlling standing, bowing & sitting actions.

Contacts for controlling the movements.

Lever controlling bowing movement.

←Driving Cables.

Geared mechanism driven by an electric motor which revolves pulley "A". This operates lever "B" causing the Robot to sit or stand.

A

B

Counter Weight.→

A Unit of the Electrical Mechanism for moving arms, hands & head.

Shades of Disney's robot entertainers.
This metal figure appeared at the 1928 London Exposition.
As part of its act it rose from its seat, bowed to the humans present,
and delivered a short speech.

speedy manufacture. This means leaving out such useless human attributes as feelings and emotions; all that remains is the capacity to work. The robots are sold as general-purpose laborers, and in due time are used as soldiers in a war. Then one of Rossum's associates finds a way to install pain and emotions in the robots. Thus endowed, the robots revolt, turn against the humans, and take over the world, virtually exterminating the human race. But now a problem arises for the robots, for it appears that they will not be able to manufacture more of their kind. At the end of the play, however, two of them fall in love. The audience is left with the suggestion that they will be the new Adam and Eve.

It is difficult to recall a literary work more in keeping with the mood of the 1920s. It gave rise to a host of works written in a similar vein, including the 1926 movie classic *Metropolis*. This film was produced by Fritz Lang, Germany's leading filmmaker, and was based on a book written by his wife, Thea Harbou. At the time, Lang's influence in Germany was immense. The premiere of one of his films would draw an audience that included virtually anyone of note in arts and letters and any number of high government officials, all in evening dress. *Metropolis* focuses on the wretched lives of the workers who live beneath a city. Its robot is a labor agitator, Maria, who assumes the appearance of a leader whom the workers trust. But Maria is really an *agent provocateur*, who incites the workers to destroy their underground world, thereby dooming themselves. In the film's climax, as the workers burn the robot at the stake, they see the figure change from a woman into a creature of metal.

R.U.R. and *Metropolis* marked the end of the prehistory of robots, and the beginning of their true history. Inevitably, Čapek's and Lang's gloomy views of the subject were bound to give rise to a rejoinder. It came from a young science-fiction writer, Isaac Asimov. In 1942, in a story in *Galaxy Science Fiction* titled "The Caves of Steel," he set forth his Three Laws of Robotics:

- A robot may not injure a human being, or through inaction allow a human being to come to harm.
- A robot must obey the orders given it by human beings except where such orders would conflict with the First Law.
- A robot must protect its own existence as long as such protection does not conflict with the First or Second Law.

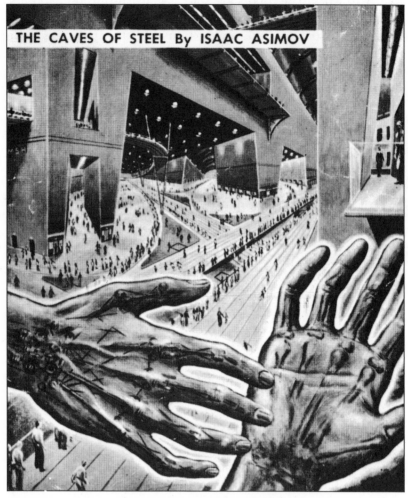

The modern concept of the robot
as an intelligent humanoid companion rather than a monster
was born in Isaac Asimov's science fiction of the 1940s,
as in his classic "The Caves of Steel."

Asimov later claimed that these laws were actually the work of John Campbell, editor of *Astounding Science Fiction* and mentor to Asimov's generation of sci-fi writers. Campbell, in turn, has always credited them to Asimov. In any event, it was Asimov who used these laws as the basis for stories. In "Evidence," which appeared in the September 1946 *Astounding*, a district attorney is running for mayor. A heckler accuses him of being a robot because no one has ever seen him eat, drink, or sleep— or for that matter prosecute anyone in court. He punches the heckler on the chin, something no robot would do under the First Law. This proves that he is human, and he is duly elected. But a question remains. Was the heckler a robot planted to serve the candidate's purposes?

Asimov's ideas did not take the literary world by storm, but over the years they came to predominate in popular attitudes. He introduced the term *robotics*, but he had no clear concept of how robots would actually be built and described them as having "positronic brains." (The positron, or negative electron, was discovered in 1932.) Had Asimov used the adjective *electronic*, he might have been on the right track. It is tribute to his foresight that his Three Laws gained rather than lost influence, once the computer was at hand. Our modern concept of the sci-fi robot remains that of Asimov as exemplified by Robby in the 1956 film *Forbidden Planet*, and by R2D2 and C3PO in the *Star Wars* trilogy. Čapek's vision has faded into a melodramatic cliché.

Two technical innovations, in the 1940s and 1950s, revived the prospect of building manlike machines, following a hiatus of a century and a half. First came automatic control (self-guiding mechanisms), developed during World War II; this led to the postwar generations of programmable robots that today are seeing increasing use in factories and industrial shops. Hard on its heels came the programmable computer. While the first simple computers served to control factory robots, the most interesting possibilities were opening up the field of artificial intelligence, with its hope that the robots of the future might truly display human characteristics.

Automatic control had gotten off to a good start prior to World War II. As early as 1922, naval engineers were discussing the automatic steering of ships, the forerunner of autopilots for aircraft. The key to this was feedback. The problem was first to

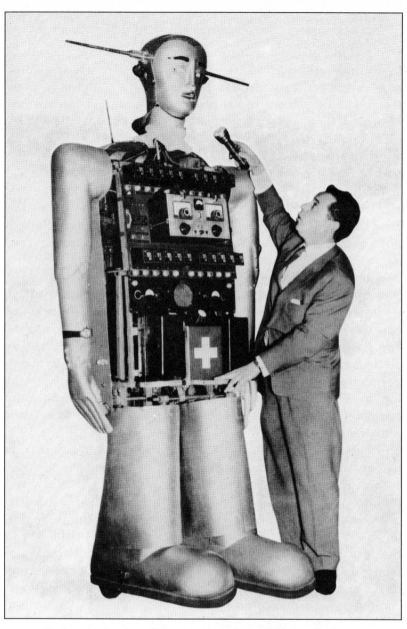

During the heyday of the robot as science-fiction cult object,
machines like Sabor, humanoid hulks of metal, were common.

Our preconceptions and expectations of robots have been
shaped by fantasy, embodied here in the forms of C3PO and R2D2.

make automatic measurements of variables such as present position, velocity, and angle; then to compare these measured values with values that the ship's captain needs to keep on course; and finally to engage a servomechanism, a self-correcting, automatic mechanism (the term was introduced in 1944), to reduce the difference to zero. For example, an autopilot might have the task of keeping an aircraft in straight and level flight, at a predetermined speed. When the aircraft wanders from this course, the resulting deviations are measured and the measurements fed back to a servocontroller. A mechanism adjusts the throttle, ailerons, rudder, and elevators to return the plane to its proper path.

World War II gave an enormous impetus to this field. At MIT, the Servomechanisms Laboratory addressed the problem of using feedback control for high-speed aiming of antiaircraft guns using radar; the radar systems came out of another MIT center, the Radiation Laboratory. General Electric pitched in with a fire-control system for the B-29 bomber. As a gunner aimed a gunsight, an electronic remote-control system swung the guns around to point at the target. In Germany, the unmanned V-1 buzz bomb featured an autopilot; the V-2 rocket had an inertial guidance system that used a gyroscope to measure its velocity and cut off fuel to its rocket engine at the proper moment. On the receiving end of these weapons, Londoners spoke of the "robot blitz." The American poet MacKinley Kantor, writing of the war, put it this way:

> And BBC re-broadcast to the States
> A tale of robot bombs. Boone City heard the bombs.
> Re-broadcast bombs are not so frightening to hear.

Five pathbreaking weapons came out of that war: radar, jet aircraft, the V-2 rocket, the B-29 heavy bomber, and the atomic bomb. All were closely associated with the development of automatic control.

Hard on the heels of these inventions came the electronic computer. In little more than a decade, it advanced from an idea in a mathematical paper to a working machine with all the basic features of computers in use today. Although a handful of brilliant Germans, working on their own, independently devised many of the key concepts and inventions, the computer as we know it sprang from wartime projects in England and America. It started in 1937, when a twenty-four-year-old British mathematician, Alan Turing, published a paper titled "On Computable Numbers." Reviving the ideas of Charles Babbage, he argued that if an automatic computing machine could be built, capable of carrying out a simple set of basic operations involving switches that would be either on or off, then this machine would be able to carry out any mathematical calculation capable of being completed in a finite number of steps.

Then, in 1939, the war broke out. Very soon, through the help of agents in Poland, the British intelligence service captured a German Enigma coding machine, used for top-secret messages. If the code could be cracked, the British would be able to read German communications and anticipate their strategy. In great secrecy, they set up a cryptanalysis project at Bletchley Park, north of London. There Turing was a leader in a group that built Colossus, a very fast vacuum-tube computer that read data from punched paper tape at a rate of five thousand characters per second and cracked the code. "I won't say that what Turing did made us win the war," said one of his close associates, "but I daresay we might have lost it without him." As early as February 1940, Turing and his associates were listening in on the Luftwaffe. For their part, the Germans never caught on—was not their code scientific?—and blissfully went on using their Enigma machines throughout the war. The British went right on, not always blissfully, reading their coded communications. It has been said that in November 1940, Winston Churchill learned that the Luftwaffe was preparing a massive air strike against Coventry. He had a clear opportunity to take defensive measures, but to do so might tip off Berlin that their secret was out. Rather than risk compromising the work of the cryptanalysts at Bletchley Park, Churchill allowed Coventry to be destroyed.

Turing's Colossus pushed the limits of its art, but it could do only one thing: break Enigma codes. In America, the first large electronic computer was ENIAC, Electronic Numerical Integrator And Calculator, built at the University of Pennsylvania. ENIAC was completed in February 1946. It is worth comparing it to one of the early microcomputers, the Fairchild F8 of 1976:

	ENIAC	F8	Comments
Size	3,000 cubic feet	0.011 cubic feet	300,000 times larger
Power	140 kilowatts	2.5 watts	56,000 times more
Read-only memory	16 K bits (relays and switches)	16 K bits	Equal amount
Random-access memory	1 K bits (flip-flop accumulators)	8 K bits	8 times less
Transistors or tubes	18,000 tubes	20,000 transistors	About the same
Resistors	70,000	None	F8 used active devices as resistors
Capacitors	10,000	2	5,000 times more
Relays and switches	7,500	None	
Add time	200 microseconds (12 digits)	150 microseconds (8 digits)	About the same
Mean time to failure	Hours	Years	More than 10,000 times less reliable
Weight	30 tons	Less than 1 pound	100,000 times more

Despite the vast differences in its technical bases, ENIAC lacked only two features of today's computers. One, it worked with the decimal system, not the binary system, which is faster and more natural for a computer. The decimal system derives from our ten fingers, and this feature was imposed on ENIAC, requiring that digits be stored as rings of vacuum tubes. The binary system is well suited to electronic components, which can be either on or off; hence a binary digit, or bit, can be stored with a single such component.

The second feature was more fundamental. ENIAC could be programmed only by wiring up a plugboard, that is, inserting individual plugs into sockets. This set up the specific electrical paths whereby ENIAC would carry out its computations. This amounted to a rigid list of commands, as unvarying as those in the cams of the old Jaquet-Droz automata. It was the insight of Princeton mathematician John von Neumann that a computer should store its program using the same electronic code used for the data it manipulated. Hence the program itself would amount to a kind of data to be manipulated. Rather than be a fixed sequence of steps, a computer program could include a lengthy repertoire of possible sequences, and its actual operations might not be predictable in advance. The computer would be able to make a calculation, compare the result with another number, and then use the comparison as a command to choose a sequence of steps to execute.

The next large computer after ENIAC, von Neumann's EDVAC of 1947, featured both binary arithmetic and program storage in an electronic memory; and so it has been ever since.

The early computers were not single devices but rooms—big rooms, full of tubes, circuits, ventilating equipment, and people. The people carried tapes, pushed buttons, lowered the room temperature; they were just as important to the computers as were the tubes and circuits. To them, these early computers were scarce and remarkable resources. Thus they had to be prepared to meet the computer on its own terms: arduously debugging their equipment, working at two in the morning to have computations ready for the next business day. All this was part of the challenge of these fascinating new machines, whose potential was limited only by the cleverness of the high priesthood who were developing their programs.

One of the highest of the priests was Claude Shannon of MIT. He had made his reputation as a graduate student, by showing how Boolean algebra, a type of formal mathematical logic, could be used to design electrical circuits. He then went on to develop many key ideas in communications theory, the formal study of how data and information are sent over a channel. (These represented major contributions to the burgeoning fields of information and control theory, which Norbert Wiener of MIT called cybernetics.) In 1950 Shannon wrote an article, "A Chess-Playing Machine." He argued that computers would be able to play good (though not perfect) chess by studying a large number of possible moves and sequences of moves, then choosing the best.

Checkers could also be played by computer and proved an easier conquest. During the 1950s, Arthur Samuel of IBM developed a checkers-playing program that would learn from experience. As in Shannon's approach to chess, Samuel's program could evaluate moves, countermoves, and the likely responses of its opponent, as if it were a human player looking several moves ahead. In addition, it kept a record of its previous games. Thus if it encountered a position on the board that it had seen before, it could choose its next move not merely by looking ahead, but by recalling what had happened the last time it was in that position. Samuel's program went from strength to strength, building an impressive body of checkers experience. In 1962 it beat Robert Nealey, a nationally known player. Nealey commented that "the computer had to make several star moves in order to get the win [and] played a perfect ending without one misstep. . . . I have not had such competition from any human being since 1954, when I lost my last game."

During the summer of 1956, some of the thinking in this area began to crystallize. John McCarthy (then at Dartmouth), along with Shannon and Marvin Minsky (then a Harvard mathematician), had sent in a proposal to the Rockefeller Foundation: "We propose that a two-month, ten-man study of artificial intelligence be carried out . . ." It was the first use of the term "artificial intelligence"; McCarthy had coined it. The study took place that summer. Arthur Samuel came, along with several others who would rise to leadership in this new field. Among them were two who were invited almost as an afterthought, Allen Newell of the Rand Corporation and Herbert Simon of

Carnegie Tech (now Carnegie-Mellon University). It was a significant afterthought. They brought with them the first working program that could reasonably be said to allow a computer to think. They called it the Logic Theorist; it could prove the theorems in the *Principia Mathematica,* the seminal work on the foundations of mathematics written by Bertrand Russell and Alfred North Whitehead. Indeed, when faced with one such theorem, the Logic Theorist devised a shorter and more satisfying proof than Russell and Whitehead had given. Simon wrote about this to Russell, who responded with delight.

Newell and Simon had worked with another Rand researcher, Cliff Shaw, in developing the Logic Theorist. These scientists quickly followed their success with the General Problem Solver. It was designed to apply certain general processes in problem-solving, including means-ends analysis and planning. Means-ends analysis meant looking at where we are, comparing it to where we want to be, and looking for ways to reduce the difference. Planning meant identifying various goals along the way, the pursuit of which could perhaps be somewhat more straightforward and bring us closer to the desired state. As Newell put it, this program had "pretensions to generality." It worked successfully on a number of tasks, including logic problems and puzzles of the type played by fans of mathematical games.

The General Problem Solver was based on the idea that one could use standard methods to solve a problem, without having much specialized knowledge about the problem itself. But this approach could go only so far. John McCarthy, who had moved to MIT, now had a different approach. He wanted to write a program called the Advice Taker, which would improve its performance by having statements given to it while it was running. McCarthy hoped the program would be able to deduce the logical consequences of this information and combine them with what it already knew. This approach proved even harder to pursue than that of Newell, Simon, and Shaw, but it had at least one significant consequence. The idea of interrupting a running program in midcourse gave rise to time-sharing, whereby a single computer could be used simultaneously by a large number of people.

Meanwhile, the first industrial robots were on their way. In the words of Joseph Engelberger, founder of the robot-building firm of Unimation Inc., "One George C. Devol propitiously turned up at a cocktail party in 1956 with a tall tale of a patent application labeled 'Programmable Article Transfer.' It was issued in 1961 as U.S. Patent 2,988,237, and good friend George went on to amass numerous other patents in robotics, to the ultimate benefit of Unimation." Engelberger set up his company with financing from two corporate executives, Norman Schafler of Condec Corporation and Champ Carry of Pullman, Inc. His first robots went out into the world in 1961.

These robots, called Unimates, featured feedback control systems linked to a computer memory. A user would operate a handheld controller to guide the machine through the desired sequence of steps. At each step, he would enter in memory the position of the Unimate. Then, in operation on its own, the robot would simply play back and execute these recorded steps. The memory would supply a desired position; servomechanisms would then shift the Unimate from wherever it was to match this specified position. The first task for the Unimate was tending diecasting machines. This common industrial process produces metal parts by injecting molten zinc or aluminum into a machined steel die. For humans it is a very unpleasant job—tedious, hot, and replete with noxious fumes. Yet it cannot be handled by the sort of conventional automation used in mass production, like turning out glass bottles by the million; most die-casting shops deal only with small batches. The new Unimate robots made it easy to set up new programs when new batches had to be produced. At the same time, when a repeat of an old batch was called for, the appropriate program could readily be called up. This flexibility made die-casting an early success story for Engelberger's industrial robots.

But these robots were a far cry from Robby of *Forbidden Planet* or Maria of *Metropolis*. They looked nothing at all like, say, the Tin Woodman of Oz, though they perhaps had a passing resemblance to his arm. It was the arm, after all, that *was* the robot. Its brains (computer) could be in a case located almost anywhere, and it had no need for legs or a body, let alone the ability to sing "If I Only Had a Heart."

Still, it was inevitable that scientists would seek to build more advanced robots, incorporating what they learned about artificial intelligence. In the mid-1960s, three such projects got under way at MIT, Stanford, and SRI International (then known as Stanford Research Institute). The Stanford and MIT robots featured an articulated arm and a TV camera. A computer received the images from the TV, seeking to extract information with which to guide the arm. The world of these robots was one of toy blocks, which could be recognized in the TV images and moved about by command. After a while, the Stanford robot graduated to the more difficult task of assembling an auto's water pump from parts scattered on a table. This task was far from trivial. The computer had to identify parts from their TV outlines. It had to know that the big nut went with the big screw and that a part put on prematurely had to be removed to accommodate another part left out in the first attempt.

The SRI robot went the others one better; it could move. It carried a TV camera and had more than a passing resemblance to R2D2 of *Star Wars*. One of the SRI scientists christened it by saying, "Hey, it shakes like hell and moves around, let's just call it Shakey." It represented an attempt to combine learning programs such as Samuel's checkers-player, pattern-recognition programs capable of deriving information from a TV image, problem-solvers related to the General Problem Solver, and programs to represent information about the outside world. In the words of one of the designers, Bertram Raphael, the hope was "to see if we could make the sum greater than the parts. Suppose you have a visual-perception capability that can give information to the problem-solver, and a problem-solver that can predict what you're likely to be looking at to help the vision system." Shakey was controlled by radio link with a minicomputer. Its vision system could pick out the dark baseboard around the bottoms of the walls in the rooms through which it moved. Also, Shakey could recognize the shapes of various boxes, as well as doorways and corners within its rooms. In one experiment, Shakey was placed in a room where there was a platform with a box on top, and a free-standing ramp some distance away. It was given the question: "Can you knock over the box?" Its problem-solver was able to decide for itself that an appropriate intermediate goal would be to push the ramp up against the platform. With that, Shakey wheeled itself up the ramp and knocked down the box.

During the 1970s, robots began to enter the workplace by the thousands. In Danbury, Connecticut, Unimation took on robots as its sole product line in 1972. (It was 1975—fourteen years after sending the first Unimates into the world—before the company turned a profit.) In 1978 its engineers introduced a smaller robot called PUMA, Programmable Universal Machine for Assembly, designed specifically to handle small parts used in assembling instruments and motors. By the end of the decade Unimation was shipping more than fifty robots a month. Only slightly less busy was Cincinnati Milacron, builder of a sophisticated robot called the T-3.

These machines cost $40,000 to $100,000 each, which averages out to about $6 per hour to operate. These robots didn't get bored, take vacations, qualify for pensions, or leave soft-drink cans rattling around inside the assembled products. They "cheerfully" worked around the clock and would accept heat, radioactivity, poisonous fumes, or loud noise, all without filing a grievance. They were at their best in dangerous jobs like handling irradiated parts of a nuclear reactor, and in stupefying jobs like auto welding or painting—both of which are best done at temperatures hotter than a human could stand. They revolutionized the auto industry. At a Chrysler plant in Detroit, fifty such robots, craning forward and emitting sparks, worked two shifts and did the work of two hundred welders who had formerly worked there. At a Ford plant in Wixom, Michigan, robots measured openings for windows, doors, and lights, working ten times faster than humans.

Robots also came to be valued outside the auto industry. At a Pratt & Whitney plant in Connecticut, ten Unimates built ceramic molds for manufacturing turbine engine blades. Using these molds helped to increase production from 50,000 to 90,000 blades per year. Moreover, the robot-made molds were so much more uniform that their blades lasted twice as long as blades molded by humans.

At the General Dynamics plant in Fort Worth, Texas, a T-3 selects drill bits from a tool rack, drills a set of holes to a tolerance of 0.005 inches, and machines the periphery of 250 kinds of parts. The robot makes parts up to five times faster than a human, with zero rejections; it cost more than $600,000 but paid for itself in eight months of operation. Moreover, other people were coming forward with new uses for robots. Unimation's

Joseph Engelberger told of getting a call from some Australians who thought robots would be just the thing for shearing sheep. He politely told them they were crazy. They explained that Australia has 14 million people but over 130 million sheep, most of which require an annual barbering. To replace their expensive sheepshearers they might need as many as 130,000 robots. That would be nearly 200 years of Engelberger's production. He suddenly saw the underlying wisdom of the idea.

All this might have raised the specter of angry workers wielding clubs to bash the sources of their unemployment. However, most industries where they were being used were strongly unionized, and the unions could ensure that robots merely replaced men who quit or were transferred, rather than forcing layoffs. Many union leaders adopted the attitude: "No firing of our people, and give us some of the goodies that come from increased productivity." As the president of a New Jersey electrical workers' local said, "If we can bring in a robot to do, say, the painting that a man does for $7.00, we can move him to another job at $7.50 an hour. We say, 'Train our people for the skilled jobs that are in today's market.' " On the shop floor, robots often went through their mechanical calisthenics not only unattended but hardly noticed by their human colleagues. The arms and boxes were, after all, just robots. When they were noticed, it was often favorably. Soon industrial observers reported that working with a robot seemed to confer status. Some of them have even inspired affection. When one machine known as Clyde the Claw broke down at a Ford plant in Chicago, its human co-workers gave it a get-well party.

A 1980 census of robots, taken by Bache Halsey Stuart Shields, Inc., showed that the United States had 3,000 of them. In Europe, West Germany led with 850, Sweden had 600, Italy 500, and France, Norway, and England each had about 200. The entire Soviet Union had only 25, and these were evidently experimental devices, but Poland had 360. Ironically, Czechoslovakia, home of Karel Čapek, was not listed. The true homeland of the robot appears to be Japan, with 10,000 in the census, more than the rest of the world combined. The Japanese government achieved this in a characteristic way, by deliberately setting out to make it easy for industrial managers to use robots. Robots were expensive, and many managers would have been reluctant to buy them without first having proven them out with on-the-job experience. The problem was how to let businesses get this experience without incurring high costs. The Japanese solution was to set up a robot-leasing company to buy the robots from the manufacturers in large numbers. User companies, able to lease robots without having to buy them, were willing to take on many robots, since they knew they could return them to the leasing company if things did not work out.

By 1980 the Japanese were producing some 7,500 robots per year, compared with 1,500 in the United States. Early in 1981 the firm of Fujitsu Fanuc opened a $38-million plant, with robots working round the clock to produce—more robots. At a Matsushita color TV plant near Osaka, about 80 percent of the parts for a set were being put in place by robots. Describing his visit to this plant, Philip Abelson, editor of *Science* magazine, wrote: "At the end of the tour, I realized that I had seen only one inspection station—at the end of the line. I asked the official accompanying me why there were no inspection stations at intermediate points. He replied that until a few months ago there had been such stations, but they never found any defects and so they were scrapped. When I returned to the United States, I tried to arrange to see a comparable plant here. I was told that none existed."

Virtually all these robots, however, amounted to no more than "blind grabbers." For instance, Unimation's PUMA could screw a light bulb into a socket, but bulb and socket had to be precisely where it was told they were. If the bulb was upside down in its tray, PUMA would contentedly try to screw it in upside down. If the socket was a quarter-inch out of position, PUMA would jam the bulb against its rim and twist away. If the bulb wasn't there, the robot would close its gripper on empty air and try to screw it in. Obviously the machine is limited, and a great deal of effort and expense have gone into making sure that parts and assemblies are aligned properly to be exactly where the robot expects to find them.

The problem of making robots smarter is one phase of the research into artificial intelligence. One topic receiving much attention has been that of giving vision to a robot. It has been easy to form an image with a TV camera, breaking a scene into small elements that are converted to numbers. The difficult part has been to interpret the scene in terms of patterns of numbers.

Even two-dimensional vision, the reading of printed text, has pushed researchers' ingenuity to the limit. In the course of an hour's reading, any of us may switch from the somewhat messy print of a newspaper to the hand lettering of the comics page, then to the clean typeface of this book. Yet we can read them all without being bothered by changes in the type style, because we carry an idea of what the letter *A*, or any other letter, looks like. But it has proven quite difficult to give this ability to a computer. The first "omnifont" reader, capable of imaging a page of common print or typescript and storing its letters or characters in machine-readable form, came out in 1976. It was the work of Ray Kurzweil, a self-educated computer genius in Cambridge, Massachusetts, who had only a B.S. degree, had never worked for a boss, and had been a successful computer entrepreneur since he was thirteen years old. Following his success, he watched bemusedly as several dozen other firms launched projects to devise their own omnifont readers. As of 1984, by his count, some fifty had tried; none had succeeded.

The trend in recent years has been to recognize that artificial intelligence must bring vast stores of knowledge to bear on a problem before it can be solved. This has represented a turnabout from the days of the General Problem Solver, whose authors argued that it would be able to solve any one of a wide range of problems by applying a few powerful methods, without knowing a lot about the background to the problem.

Understanding natural language, allowing a person to talk with a computer in ordinary English, is an example of one problem that calls for a vast knowledge base. If a computer could understand English, it could do a good job of translating languages. In 1949, Warren Weaver of the Rockefeller Foundation proposed that machine translation could be done using the codebreaking methods that had come from the wartime work of Turing and others. As Weaver put it, "When I look at an article in Russian, I say, 'This is really written in English, but it has been coded in some strange symbols. I will now proceed to decode.' " Legend has it that one early researcher asked his computer to translate, "The spirit is willing but the flesh is weak," first into Russian and back into English, and that the result read, "The vodka is agreeable but the meat is rotten." (Actually, the early machine-translation programs were not even that good.)

Another early language program was ELIZA, written by MIT's Joseph Weitzenbaum in 1966. It took its name from the famous Miss Eliza Doolittle of *My Fair Lady*, who had to be taught proper English. ELIZA bypassed any real language understanding, relying instead on a clever system of fixed patterns of response. If someone used the word *mother*, ELIZA would reply with a stock response, such as, "Tell me more about your mother." Weitzenbaum compared this to the polite chatter of a cocktail party, where one can keep a conversation going with such statements as "I see" or "Please tell me more." Even so, a number of people clamored for the chance to talk by teletype with ELIZA, and occasionally even a sophisticated computer expert could be fooled.

One Saturday morning in 1968, for example, one vice president at the high-tech firm of Bolt, Beranek and Newman in Cambridge, Massachusetts, arrived at the computer room. He found a note on the teletype: *Type all messages into this teletype, and end them with a period.* The v.p. thought this meant the computer was linked to a terminal in the home of one of his top programmers, Daniel Bobrow. Actually, it was running a version of ELIZA. The v.p. then started what he thought was a teletype conversation with Bobrow:

      V.P. Do you think that I can use the computer this morning?
    ELIZA Why do you ask?
      V.P. I might be able to make some additional telcomp sales.
    ELIZA Why aren't you sure?
      V.P. My prospective customers have never seen a demonstration of the system.
    ELIZA Does this mean very much to you?
      V.P. Of course it does.
    ELIZA You seem very positive about that.
      V.P. This is ridiculous.
    ELIZA What do you mean this is ridiculous?
      V.P. Is there any reason why I can't use the computer now?
    ELIZA Tell me more about your prospective customers.
      V.P. Please dial me up on 491-1850

After that last statement, the v.p. left out the period, and the computer didn't answer him. This infuriated the v.p., who thought

Bobrow was playing games with him. He called Bobrow at home and said:

> V.P. Why are you being so snotty to me?
> BOBROW What do you mean, why am I being snotty to you?

The v.p. angrily read him the dialogue and couldn't get any response from Bobrow but laughter. It took a while for him to convince the v.p. that it really was ELIZA.

Since then other programs have been written to deal with natural language to at least a limited extent. In the mid-1970s, at Bell Labs, an experimental system could not only listen to a speaker but could also process information extracted from his speech. It then went on to compute an appropriate response, encode the response in a proper English sentence, and deliver a reply using synthesized speech. The system acted as an airline ticket agent, with a vocabulary of 127 words. This is roughly the level of a student in his first week of studying a foreign language, where the lessons feature a highly limited vocabulary keyed to just this kind of stock situation—a brief conversation at a ticket counter. The gap between these elementary lessons and the ability to read Sartre or Solzhenitsyn in the original compares to how far computers have to go before they can become truly fluent.

Ironically, although it is difficult to codify our commonsense understanding of the world, at least it has been possible to codify a variety of limited and specialized bodies of knowledge. The resulting programs were the expert systems, which attempt to deal with the highly technical worlds of experts in medicine and technology. It turns out that to mimic a highly trained expert is much simpler than duplicating "everyday" language. Expertise proves to be easier *because* it is specialized. It focuses on narrow classes of problems and calls for limited amounts of knowledge about the world.

Here are a few of these specialized programs: DENDRAL interprets data from chemists' instruments and gives advice on the structure of unknown compounds. MACSYMA, for mathematicians, carries out the complex symbol manipulations in calculus and higher math. PROSPECTOR is for geologists; in 1982 it identified the location of a previously unknown deposit of molybdenum. MYCIN prescribes treatments for meningitis and infections of the blood, and can help in diagnosis as well; and a similar diagnostician, INTERNIST, deals with problems in internal medicine. In one evaluation of forty-three cases taken from the *New England Journal of Medicine*, INTERNIST made the correct diagnosis twenty-five times, compared with twenty-eight times for the physicians who were caring for the patients.

It would probably be easier to develop an expert system capable of answering questions about nuclear reactor design than to devise a system capable of sending Shakey to the store to buy a loaf of bread. And with this, the fields of artificial intelligence and robotics have come face to face with an issue that goes back to their roots in antiquity. We find the ancient Greeks asserting that the form and activities of men were praiseworthy, deserving of being reproduced in works of artificers that might be cleverly wrought. These same philosophers gave us a highly influential ordering of the relative merits of human mental activity. The highest work was that of a (surprise, surprise) philosopher. High accolades would go to those who could prove theorems in mathematics, master the arcana of a physician, or excel at such games as chess. Lower by far would be artisans and technicians. The activities of ordinary life, including language and vision, were beneath mention.

It is hard to overestimate the power of these views. Even today, a wool-gathering professor generally has a higher social status than a man who does something useful, such as rebuilding auto transmissions. Indeed, this point of view delayed the rise of the experimental method in science by encouraging the attitude that the truths of nature could be found entirely by logical reasoning. Although we no longer hold to this viewpoint with anything like the thoroughness of the medieval scholars, still its influence remains evident. The reason is simple. For twenty-five hundred years we had no alternative form of intelligence to serve as a standard of comparison.

From the perspective of artificial intelligence, however, we now can say that this philosophy is totally wrong; indeed, it has things backward. For computers, the easiest task is to prove theorems and carry out solutions in higher math. Slightly harder is the playing of games at championship level. A bit harder than that is the expertise of a professional internist, geologist, or

chemist, or—and this has also been put into an expert system—
the experience of a diesel locomotive repairman. Harder still is
the ability to use ordinary language. And most difficult of all
are such matters as a commonsense understanding of the world,
and the ability to make sense of what is seen. At the highest
level of achievement is the sort of task that would have been an
everyday matter to our prehuman ancestors. In the words of
C. R. Carpenter of Penn State, a longtime observer of monkeys
in the wild:

> You are a monkey, and you're running along a path past
> a rock and unexpectedly meet another animal face to face.
> Now, before you know whether to attack it, to flee it, or
> to ignore it, you must make a series of decisions.
> Is it monkey or nonmonkey?
> If nonmonkey, is it promonkey or antimonkey?
> If monkey, is it male or female?
> If female, is she interested?
> If male, is it adult or juvenile?
> If adult, is it of my group or some other?
> If it is of my group, then what is its rank, above or
> below me?
> You have about one-fifth of a second to make all these
> decisions, or you could be attacked.

In the mid-1980s, then, while robots and artificial intelligence
have found their uses in specialized niches, they cannot be said
to have influenced our world in anything like the manner of
their parent, the computer. Yet even with their limited influence,
they have already given us something priceless. Through these
sciences, we are looking upon the machine that is like a human,
and we see more clearly than ever before—ourselves.

Robotics and Common Sense
Philip E. Agre

To exhibit common sense a robot must be able to manipulate models of its world, reason by analogy, carry out commonly useful lines of reasoning automatically, and develop enough of a "Self" to sensibly modify its behavior. An architecture for the mind of such a robot must be able to reflect the complexity of the world without being overwhelmed by it.

A woman I met on the subway understood right away why it's interesting to try to build a robot that makes breakfast.

"Oh," she said, "*I'm* a robot when *I* make breakfast."

Present applications of robots are true to the stereotype of the robot: routine in the extreme, the same stiff motions over and over, with little or no variation. We might expect making breakfast to be an interesting next task for robots because, at least for Americans, it is the stereotype of a mindless routine among our day-to-day activities. Making breakfast *feels* automatic, a rote skill to be mindlessly cranked through every morning, but that feeling is an illusion. Most of us can make breakfast and deal sensibly with the innumerable mundane contingencies of milk shortages, burned toast, dropped spoons, and absurdly tightly glued cereal packages. And human breakfast-makers who drop their spoons just pick them up, clean them, and continue on, but industrial robots that drop their screwdrivers are dumbfounded. This is the paradox of common sense: Actions that feel

71

automatic are actually backed up by a great deal of understanding of why those actions are the right ones. In other words, human breakfast-makers understand what they are doing and present-day industrial robots don't.

The robots of science fiction are still fiction, but robotics is far enough along that we don't have to leave to the imagination what robots will be able to do, and what it will take to make them do it. To guess how far off some robot ability is, we must first understand in detail what that ability *is*. That done, we can figure out what that ability might amount to—both in people and in robots. Intuition says that, among intellectual feats, playing chess is hard and making breakfast is easy. But intuition is wrong. AI researchers have found that some stereotypically intellectual tasks, like playing chess and diagnosing certain illnesses, are remarkably simple for a machine to perform reasonably well, once the tasks themselves are understood. Machines can already make medical diagnoses and play good chess.

Playing chess is easy, but making breakfast is enormously complicated. This complexity stares us in the face every morning, and yet it is invisible. The things machines now do best involve carefully defined and rigidly controlled little worlds, like the world of a chessboard or the domain of an assembly line. This is the *opposite* of common sense, which is the skill of dealing sensibly with the messiness of the world of everyday life. Rather than reflecting a single insight, or algorithm, common sense is made up of a lot of little ideas and simple skills. That any bit of common sense looks obvious by itself often masks the astounding complexity of common sense as a whole and the extraordinary difficulty of acquiring it. To understand common sense, we have to look at it *as a whole*. I propose to take a close look at the common sense of two parts of the everyday lives of ordinary people, making breakfast, and getting along with other people. Afterward, I will speculate about how we might use what we've learned about common sense in building a robot that has it.

What does your understanding of the process of making breakfast consist of? In making breakfast, you think about using your hands to work with tools, you think about the things (food) you will be manipulating with your tools, and you think about the physical processes (mixing, frying, boiling) that those manipu-

Although Wolfgang von Kempelen's Chess Player was a certified hoax,
the first glimmerings of machine intelligence were to surface
with skilled but beatable machine chess players.

lations will set in motion. Using precise theories of physics and chemistry for this is much too complex and not really necessary (Do you need to know what is going on when your cornflakes are getting soggy?). How does the surface tension of liquids work? *I* certainly don't know. It suffices to know that these things happen, and to have a rough understanding of their consequences in the few situations in which they matter.

The average breakfast-maker avoids thinking about the full complexity of breakfast-making by choosing a simplified *model* of each bit of the task. A model is just whatever intuitions you have about a bit of the world that let you guess what's going to happen in it. A good model captures the circumstances of each bit of your task in just enough detail to suggest a good-enough plan for carrying it out. Models can often be visualized and set down as diagrams. Some are more detailed than others, like a model of friction that distinguishes between static friction and sliding friction, as opposed to one that doesn't. The theories of physics and chemistry are the most detailed models of all. Simple models are easier to make and use than complex ones, but they often make wrong predictions or none at all. University of Rochester AI researcher Patrick Hayes calls the study of models of the physical world *naive physics*.

Imagine that you're trying to figure out for the first time how to eat scrambled eggs with a fork. You can think of the business end of your fork as having a number of different physical structures, depending on what you intend to do with it. For cutting your food, you're better off thinking of it as a sharp blade. For scooping the food, it's a thin, rectangular plate. For stabbing the food, it's a collection of pointed spears. For mashing the food, it's a sort of grating. It's the same fork the whole time, but its real shape is complicated. However, it's a waste of time to think about the fork's tines while scooping with it or to think about its edges while stabbing with it; a good model suppresses irrelevant details and lets you concentrate on the important ones.

Much of the common sense of breakfast-making is in knowing when to use which simplified models. Most of the thoughts our average breakfast-maker ever thinks, in fact, can be phrased in terms of perhaps a couple of hundred *generic models*. Generic models are simple models that serve as the building blocks of the more elaborate models of real tasks. Blades, plates, spears, and gratings are generic models of the physical structures of objects; other generic models concern physical processes—plans, substances, operations on substances (like mixing and carrying), and so on. Several generic models can be applicable to one situation, and the same generic model can show up in a wide variety of tasks. By knowing the generic models and how to use them, you can (and probably do) think your way through breakfast without solving a single equation.

Each model of an object includes a plan for using it. Blades are good for cutting, flat thin plates for scooping, pointed spears for stabbing, and gratings for mashing and straining. A good rule is to use the most convenient model to make a plan, and use the least convenient model to predict what might go wrong with it. When you try to scoop up the last of your eggs, the business end of your fork might behave not like a thin plate but like a surface whose edge pushes the eggs away. If this happens, you can patch your plan by blocking the eggs with a knife to keep them from being pushed away. You might also choose a different model of the fork and try stabbing with its tines. In desperation, you can use your hands; because human hands can be made to fit so many different models, one of them is sure to suggest a plan that will work.

The same models of the physical world that help you to move around in it also help you interpret your perceptions of it. Your models of objects and processes make assertions about them that you can use in making deductions. When your knife won't move any farther into a peach, you can assume it has hit the stone. When the kettle is whistling, the water is boiling. When you hear a cracking sound, it may have been an egg that rolled off the countertop.

Some generic models, like the different models of a fork's structure, are specific to thinking about the physical world. Some models, though, are much more widely applicable and can be found underlying all kinds of reasoning. MIT AI researcher David Chapman calls these *cognitive clichés*. Consider, for example, the idea of a *resource*. (Milk, money, time, train tickets, and slices of banana on your cornflakes can all be resources.) Some activities consume resources, either continuously or in chunks; if you keep consuming a resource you can run out of it entirely; and if you're in danger of running out of a resource you may want to conserve it. Something you figure out about resources in one context (make sure you always have some extra milk on hand, just in case) might be as useful for other resources (money, for example, or the time allocated for a project). Cognitive clichés are the simplest and most powerful of models, so much so that whole branches of mathematics are devoted to investigating their properties.

To give you some idea of the range and properties of generic models, here is a rough sketch of how a breakfast-making robot might use them to understand why there's a limit to the number of eggs it can cook in an omelette in a given pan. (The names of the various pieces of generic models it might use appear in italics.) This story isn't physically accurate in every detail, but rather a series of well-chosen approximations.

The robot must first beat some eggs. This turns a *finite set* of four eggs into a *liquid object*, which the robot then pours into a pan. With the change in form of the egg, the models provide a change in the form of the question. *Have we got too many eggs on the countertop?* becomes *Have we got too much egg in the pan?* The two questions are equivalent, even though eggs in shells are measured by counting and liquid egg is measured by volume. (This isn't obvious to small children, or to robots.) Since the count of four eggs as too many hasn't got much to do with the volume of egg-stuff in the pan, the first question can be answered only by answering the second.

More factors to consider. The stove's burner provides *energy*. Because its flame is in contact with the pan and the pan is in contact with the egg, a *process* of heat *flow* exists between the flame and the pan and the egg. Whenever something is flowing into an object, the corresponding *state variable* of that object increases. So, the robot infers, the longer the egg stays in the pan, the hotter it gets. As the egg gets hotter, it passes through three *phases:* first liquid, then congealed, and, finally, burned.

In thinking about eggs, the robot will find what is hard is that the simplest models make wrong predictions. Each time this happens, the robot will have to convert its understanding of omelette-making into a new, more complex model. In the simplest model, there is just one object made of egg, and it occupies the whole pan. As heat flows into this egg-object, it will change from liquid to congealed to burned. This model predicts that no

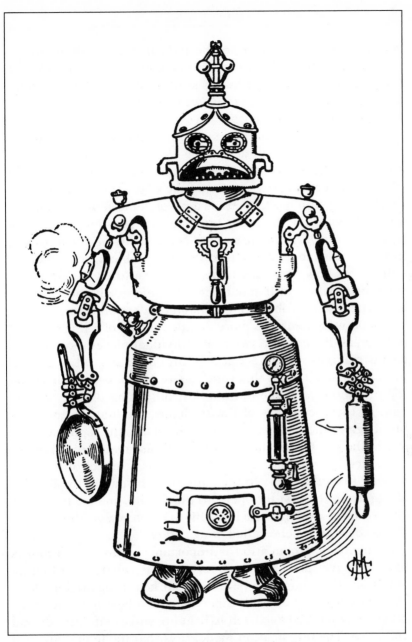

The Breakfast-Maker Robot of fantasy.
In real life, the morning challenge does require
a surprising amount of intelligence.

matter how much egg is in the pan, it will be all liquid or congealed or burned. But when black smoke appears from a panful of half-cooked egg that is still too slimy to eat, the robot will discover that this isn't always true.

A failed model can suggest ways of choosing a new model. In this system (frying pan and its contents on a stove), all the processes operate *uniformly* across the *plane* of the pan's surface— a good pan heats evenly. This suggests that the robot make a model of a vertical *cross-section* of the egg sitting in the pan. In this cross-section model there is a bit of flame, a bit of pan, and a line of egg. Now the robot translates its question, from *Have we got too much egg in the pan?* to *Is the egg in the pan too deep?* But even this model doesn't explain the problem, since there is still a heat flow from the flame to the pan to the single linear egg-object, which still should be all liquid or all congealed or all burned. So the robot decides that it needs another new model.

A good way to make a more detailed model of a linear object is to break it up into a *chain* of individual pieces. So now the robot imagines—all in a stack—a bit of flame, a bit of pan on top of the flame, and, this time, many bits of egg, each bit on top of another and the bottommost bit sitting directly on the pan. In the vocabulary of linear chains, the first, bottommost bit of egg in the pan is the *front* of the chain, the topmost bit of egg is the *end* of the chain, and the phrase *on-top-of* expresses the *successor relation*. Now the robot has a model of the heat flowing from the flame to the pan, from the pan to the first bit of egg, from the first bit of egg to the second, and (so the linear chain model says) from each egg bit to its successor; this process is the cognitive cliché called *propagation*.

Recognizing propagation is important, because the robot is likely to have run into other propagations; whatever it figured out about them ought to be true about heat propagating along chains of egg-stuff as well. Since there is a heat flow into each bit of egg, each bit of egg, left to heat up, will eventually congeal and burn. The propagation model says that the temperature of each bit of egg in the chain will be higher than that of its successor, and so each bit of egg will congeal or burn before its successor. This will suggest that there is an entity, a congealing *transition*, that propagates along the chain. It starts at the bottom of the egg and eventually reaches the top. So too will another entity, the burning transition. The longer the chain, the longer an entity takes to propagate. The breakfast-maker can finally conclude that if the chain is long enough, then the congealing entity will still be moving toward the top as the burning starts from the bottom. In other words, if you have too many eggs in the pan then the bottom will burn before the top has congealed. You can alleviate the problem by turning the heat way down and waiting half an hour; this is called a *frittata*. Or you can distribute the heat more evenly over the eggs by stirring them; this is called *scrambled eggs*.

Most of the reasoning about models that a robot will have to do in the course of a day won't be this complex, but the procedure is always the same. Start with a simple model. When the model makes an incorrect prediction, use what you know about generic models to build a new model. Each time you move from one model to another, the models themselves will tell you how to translate both what you know and what you want to know into the new model. The new model will suggest some new ways of thinking about what you *do* know; in this way you might find out what you *want* to know. When you succeed you will have constructed a line of reasoning that can apply easily the next time you need it, whether the context is making pancakes or a steak, or understanding why you have to stir just about every-thing you're heating in a pan. In these cases the particulars are different, but the form of reasoning remains the same.

Whether you learned to make breakfast by watching other people do it, by figuring it out from scratch, or by some of both, the sort of understanding I have just described underlies your breakfast-making routine. No doubt there was room for im-provement in your first breakfast-making routine, and you had to deal with a lot of specific problems: How can you reliably tell how much milk to put on your cereal? How can you minimize the risk of getting eggshell in with your eggs? Do you have to use a knife to slice a banana? How can you avoid the teapot effect, whereby a liquid defies gravity and dribbles from the outside of the container it is being poured from? Do you have to make a fool of yourself to retrieve the prize from the bottom of a new box of cereal? Can you avoid the unpleasantness of eating the dust at the bottom of an old box of cereal?

The back of your mind puts a lot of energy into recognizing

solutions to the problems of your breakfast-making routine, even when those solutions are hidden. Here again the power of suggestion of generic models can help. Imagine that you had wondered if there was a way of slicing the banana for your cornflakes without having to clean another utensil. Your perceptual systems are always trying to apply the generic models to whatever is at hand. Thus, as your spoon passes through the surface of the pile of cereal in your bowl you might notice that it can also be thought of as *cutting* the cereal. Then you might arrive at a novel observation: *What use is it to cut the cereal with a spoon?* (The back of your mind is always ready for questions like these.) *How about cutting the banana with the spoon?* Many people have discovered this trick—and many haven't.

Part of common sense, then, is noticing opportunities to act more sensibly in the future and changing your plans accordingly. This ability to think about and modify your own mind is *introspection*. Other things you learn about changing your routine plans involve making one action serve several purposes, or rearranging the steps to allow several processes to proceed at once. Take another part of making breakfast, brewing tea. The steps involved in brewing tea are:

- putting a tea bag in a cup
- pouring hot water into the cup
- waiting a few minutes for the tea to steep
- throwing out the used tea bag

It is natural to put the water on to boil after setting out the cup and tea bag. But after doing it this way a few times, the introspective tea-maker will notice that it would be better to put the kettle on to boil first, since there is plenty of time to get out cup and tea bag while the water heats up. Having learned this trick, you might further break down the step of heating the water by figuring out that it is best to turn on your electric stove before filling the kettle rather than after. Though it might be possible in principle to think up all desirable improvements to a plan ahead of time, in practice it is difficult and is often a waste of time. Even so, you can learn to think of particular improvement tricks by learning to recognize the situations where each is applicable.

In ways like this, your plans evolve over time. On a larger scale, the plans of a culture also evolve, and evolving along with a culture's plans are the tools it uses. The understanding encoded in tools is part of a cultural background of assumptions you can make without thinking. You regularly assume that there are no holes in the pots and pans, that there is some salt available, that the knives are reasonably sharp, that the utensils are in a waist-high drawer in the vicinity of the sink, and that there are no hidden edges to cut yourself on. These assumptions are part of the kitchen as an institution; by making convenient models true, they let you count on simple plans working right. In this sense you have a sort of contract with your kitchen. The details of that contract are part of our culture, the accumulated wisdom about how a kitchen should be organized that is handed down from generation to generation. Since everyone relies on the kitchen to keep its contract, details of that contract in turn become matters of politeness binding on everyone who uses the kitchen. For instance:

If you don't clean it up then someone else will have to.

Because it is the first cultural institution with which most children come in contact, the kitchen is important in the study of the development of a child's reasoning about social abstractions. A child hasn't got the knowledge or the logical ability to figure out, or even entirely understand, all of the reasons that one must keep the kitchen a certain way. A child who leaves a mess or doesn't put something away is sometimes lectured and sometimes punished on the institutional reasons that what he or she did was wrong.

This confluence of logic and force sets the pattern for later, often less benign, encounters with institutions and their ideologies. The process by which a person comes to believe in social abstractions, defines herself in terms of them, and decides what to do because of them is what sociologists Thomas Berger and Robert Luckmann call the *social construction of reality*. Any robot that is going to operate outside of a highly constrained environment like that of a factory will have to understand a great deal about how to get along with others, both robots and people. Much of this understanding is handed down through the culture in the form of customs and etiquette; even so, getting along with others requires us to use common sense.

The extent to which it is already possible for a machine to "understand" the complexities of social interactions is indicated by the work of UCLA AI researcher Michael Dyer. Dyer wrote a program called BORIS that used its understanding of human emotions and of common kinds of social interactions to do a quite convincing job of understanding and answering questions about the following rather complicated story:

> Richard hadn't heard from his college roommate Paul for years. Richard had borrowed money from Paul which was never paid back, but now he had no idea where to find his old friend. When a letter finally arrived from San Francisco, Richard was anxious to find out how Paul was.
>
> Unfortunately, the news was not good. Paul's wife Sarah wanted a divorce. She also wanted the car, the house, the children, and alimony. Paul wanted the divorce, but he didn't want to see Sarah walk off with everything he had. His salary from the state school system was very small. Not knowing who to turn to, he was hoping for a favor from the only lawyer he knew. Paul gave his home phone number in case Richard felt he could help.
>
> Richard eagerly picked up the phone and dialed. After a brief conversation, Paul agreed to have lunch with him the next day. He sounded extremely relieved and grateful.
>
> The next day, as Richard was driving to the restaurant, he barely avoided hitting an old man on the street. He felt extremely upset by the incident, and had three drinks at the restaurant. When Paul arrived, Richard was fairly drunk. After the food came, Richard spilled a cup of coffee on Paul. Paul seemed very annoyed by this, so Richard offered to drive him home for a change of clothes.
>
> When Paul walked into the bedroom and found Sarah with another man he nearly had a heart attack. Then he realized what a blessing it was. With Richard there as a witness, Sarah's divorce case was shot. Richard congratulated Paul and suggested that they celebrate at dinner. Paul was eager to comply.

Dyer's program demonstrated its understanding of this story by providing correct answers to questions like these:

Why didn't Richard pay Paul back?
Richard did not know where Paul was.
Why did Paul write to Richard?
Paul wanted Richard to be his lawyer.
How did Paul feel when Richard called?
*Paul was happy because Richard agreed to be Paul's
lawyer.*
Why did Richard get drunk?
*Richard was upset about almost running over the old
man.*
How did Paul feel [when he got home]?
Paul was surprised.
Why did Richard congratulate Paul?
Paul won the divorce case.

Here is how Dyer's program works. Dyer noticed that many
stories about social interactions turn on the success or, more
commonly, the failure of someone's plans. The program locates
the characters' successes and failures by recognizing words in
the stories that indicate their emotions; positive emotions indicate successes and negative emotions indicate failures. Since
one character's success is often another's failure, the program
represents the story in terms of common ways in which plans
fail. Dyer calls these representations TAUs, for *Thematic Abstraction Units*. TAUs are often like adages one finds at the ends of
fables.
     Thematic Abstraction Units are the generic models of the so-
cial world. They capture the essences of situations and plans,
leaving out irrelevant details. Because the laws of social processes aren't nearly so tractable as the laws of physical processes,
planning in a world full of other people is much more complicated than planning in the kitchen. Planning in both worlds,
though, is based on the understanding of the deeper structure
of the world implicit in the generic models you use to represent
it. These models let you classify situations, reason by analogy,
interpret your perceptions, and build the simple and rough plans
from which your eventual finished plans derive, whether they're
for making breakfast or for observing social interactions. So in
the divorce story, the program flags:

—TAU-DIRE-STRAITS, when Richard proved himself a
  friend indeed
—TAU-CLOSE-CALL, when Richard barely missed the
  old man
—TAU-MISTAKE, when Richard spilled coffee on Paul
—TAU-RED-HANDED, when Sarah was caught by Paul
  and Richard
—TAU-HIDDEN-BLESSING, when a silver lining was
  found in spilled coffee

The ability of TAUs to capture the essence of a story can be
important in noticing analogies. One way to tell that the following stories are analogous is to see that they are both instances
of TAU-HYPOCRISY:

> Mark always complained about how unfair it was to oth-
> ers in the class when someone cheated on exams. When
> his physics class had their next exam, Mark "checked"
> his answers with those of the person next to him.

> In a lengthy interview, Reverend R severely criticized
> President Carter for having "denigrated the office of pres-
> ident" and "legitimized pornography" by agreeing to be
> interviewed in *Playboy* magazine. The interview with
> Reverend R appeared in *Penthouse* magazine.

Noticing instances of a TAU in a story often allows inferences
to be drawn as to the reasoning and intentions of the characters.
For example, upon realizing that having coffee spilled on one's
clothes can make one uncomfortable, the program invokes TAU-
MISTAKE, which suggests that Richard is going to feel guilty
and offer to make up for it. This allows the program to offer an
interpretation of Richard's offer to drive Paul home. (The program doesn't wonder why Paul accepted a car ride from a drunk
man, though, and a good thing, too, because it is not good at
understanding what people decide when there are reasons to go
either way.)

What this all means is that if we are going to build robots
that can get along with others, we must first understand what
getting along with others involves. Much of what you know about
social interactions comes in the form of the little rituals of daily
life, like exchanging good mornings with acquaintances, dealing

with waiters, standing just so far away from someone you're talking with, passing on the left, and saying *uh-huh* every few seconds on the phone to assure someone you're still listening. Although these rituals require little thought and feel automatic enough, once again we are faced with the paradox of common sense. Even when you learn about social interactions through more or less formalized rules of etiquette, you eventually acquire enough of an understanding of why they exist to enable you to recover from awkward situations and deal with exceptions without undue perplexity.

Consider the process of initiating a conversation. First you get the attention of the other party, establish eye contact, show you know who they are, and wait for some acknowledgment, which should come quickly. For example, you might say, "Hey, Dave," with a particular inflection that indicates you want to start a conversation. Or you say, "Dave?," with a different inflection if you mean to start off with a question. There are subtle conventions for signaling how long a conversation you have in mind; one might stand farther away to start a shorter conversation or say, "Let's talk," for a longer one.

Once the terms of a conversation have been established, the participants turn to one another and the initiator begins talking. If the conversation is going to last very long, the participants also place themselves an appropriate distance apart. This distance varies among cultures; three feet is common in the United States (more if there's a great difference in the speakers' heights). Often, about twenty seconds into the conversation, the two parties will exchange signs to assure one another that neither is itching to walk away. Thus, one party or the other changes position, by shifting weight to the other foot or by taking a step in one direction or another. About five seconds later, the other party reciprocates. These rules are all instances of TAU-IMPLICIT-CONTRACT. A person who stands too far away during a conversation, or uses an unconventional gesture, will be misinterpreted or thought distant or impolite—or just peculiar—and for good reason. None of this is the stuff of formal rules or etiquette books; everyone has had to figure it out for himself at some point, even though most people have forgotten that they're even doing it.

The details of most of life's little rituals of consideration are motivated by the simple desire to pursue your ends without unintentionally offending anyone. Because there are many ends you can have and many ways to offend others, you need different reasoning to arrive at these details. Even though they're obvious in retrospect, rules like these are hard to invent:

- Talk quietly around sleeping people.
- Don't talk about a party around someone who wasn't invited.
- When turning a hallway corner, don't cut it too close.
- Avoid blocking driveways when stopping for a red light.

By the time you have to learn them, most of the rituals of daily life have already been established as the customs of your culture. The fact that others are using them, and expect you to use them too, is an extra motivation to see the logic in them. A shared cultural understanding of these rituals allows them to change when their determinants change. Entirely new situations call for entirely new customs. Some early button-controlled elevators went to floors in the same order that the buttons were pushed. People who used these elevators quickly realized that it was impolite to push a button for a distant floor before waiting to see if others perhaps wanted closer floors.

The arrival of the first commonsensical robots will call for the formation of new customs. Because it is so hard to anticipate all the situations that require people and robots to decide how to act toward one another, it won't be possible to legislate most of them. They will have to evolve by the same natural process by which human customs develop. To understand this, imagine a blind person coming to work in your office or live in your home. How exactly do you get her attention to start a conversation? Should you signal her if you're holding a door for her? Can she tell if it's light or dark outside? Is it polite to use strongly visual metaphors in explaining things to her? How much help does she need with a menu? The questions will spring on you unexpectedly when the realization hits that the customs that apply to sighted people won't do. Once you work out the correct rituals for each situation, you'll be able to feel comfortable that you'll always know what to do without causing embarrassment or confusion or injury.

The rituals for interactions between robots and people are

likely to be quite different from the ones to which we're accustomed. Imagine that a new robot has arrived to work in your office or live in your home. It is likely to have a large number of attributes that will require special rituals for dealing with it. It can't see very well, or understand speech. It's clumsy (TAU-INCOMPETENT). It needs to be plugged in occasionally (TAU-DIRE-STRAITS). It can be hard to pass in the hallway if it's not careful to stay to one side (TAU-UNCOORDINATED-PLANS). Its owner won't be happy if you're mean to it (TAU-RETALIATION). It can tell different people apart but has a hard time telling if you're looking at it. It doesn't know very much, but it's fairly smart, in a strange sort of way. It likes to explore and has lots of questions. Yet, unlike a blind person, it is unlikely to qualify as fully sentient and worthy of respect until it proves itself.

By knowing how to use TAUs, a robot might be able to participate in coming up with customs to govern the interactions of daily life. These customs will have to take the capabilities of robots into account, or nobody will be able to get them to do anything. Everyone will most likely feel clumsy and frustrated at first (TAU-INEXPERIENCED). Someone will point out to the robot that its noisy motors are likely to disturb meetings. Everyone will learn that a particular tone of voice will get the robot to go away, and another will get its attention. When someone has to squeeze by it in the hallway (TAU-CLOSE-CALL), the robot will apologize (TAU-APPEASEMENT) and resolve to stay to one side in the future (TAU-PRECAUTION). With time, everyone—people and machines alike—will begin to understand what the conventions are and will count on everyone else following them (TAU-ROLE-REVERSAL). This, in turn, will cause new customs to come into being, because the more predictable the world is the more elaborate plans you can make.

How should one go about designing a mind for a common-sensical robot? Simply copying the workings of the human mind won't do, for two reasons. First, we hardly know anything about how the human mind works. Secondly, even if we did, there is no reason to believe that we could duplicate in silicon what nature has built in protoplasm. And yet, we can derive a *little* inspiration by looking at the ways that people make breakfast and get along with their fellows. Robotics and psychology can inform one another, but the process is subtle. Suppose we view the human mind as having been designed by an engineer (whether evolution or God or both). Roboticists are trying to apply the principles of engineering to the design of the minds of robots. The things we have learned about common sense can offer some insight about how we might design what computer engineers would call an *architecture* for the mind. An architecture for the mind would explain what sorts of things are stored in the mind, what happens to them once they get there, and how they enter into the mind's constant decisions about what best to do next.

The human mind is widely thought to be organized in two parts, the peripheral systems and the central systems. The *peripheral systems* include the parts of the brain that process the information coming in from the senses and the commands going out from the brain to the body. Since the problems that the peripheral systems have to solve—like seeing and hearing—have been around for millions of years, specialized neural hardware has evolved to solve them. The *central systems* are the parts of the brain that think about playing chess and diagnosing leukemia and cashing checks, tasks that nobody had to perform until more recent times. Proponents of the distinction between peripheral and central systems, such as MIT philosopher Jerry Fodor, imagine the central systems to be much more general than the peripheral systems. The central systems concern themselves not with visual images and limb motions but with abstractions, models, and plans, used in reasoning about games or diseases or banks.

Some problems the mind has to solve are especially well suited for solution by the specialized pieces of neural hardware that make up the peripheral systems. Consider the brain's first few stages of visual processing, which find important features and textures in the image on the retina. Because that image is two-dimensional, it's natural to process it with the flat sheets of neurons of which much of the brain is made. Current computer technology suggests some parallels between the human brain and the likely design of robot brains. For example, it turns out to be quite useful to build electronic circuits with two-dimensional organizations on silicon chips and printed-circuit boards that mirror the flat sheets of the brain's cortical neurons. Consequently, we can expect the design for many of the peripheral systems of robot brains to be the same as the peripheral systems of human brains.

So much for peripheral systems; what can our understanding of the problem of using common sense tell us about the architecture of the central systems? Recall the paradox of common sense: Routine actions that feel automatic in their execution are based on understanding why those actions are appropriate. Recall also the role of generic models in this understanding. Imagine, then, that every model in your mind has associated with it a collection of *canned thoughts*, thoughts you think every time you apply that model; so for example every time you discover a new insight about hypocrisy, you store the thought alongside TAU-HYPOCRISY. By doing this you arrange for that insight to occur to you each time you come across an instance of hypocrisy.

A canned thought is like a map of a trail through a dense forest, with each step corresponding to a bit of reasoning from your premises. The first trip through an unfamiliar forest can be a lot of work. All the paths look pretty much alike, and if AI has learned anything it's that most of the paths head off in pointless directions or result in dead ends. When you do find a path between a common set of premises and a useful conclusion, you can map it so that the next time you find yourself at the edge of the forest corresponding to those premises, your map will let you move quickly past the dead ends to the desired conclusion. If the premises describe a familiar problem and the conclusion describes how to solve it, then you'll reach the solution effortlessly from then on.

A mental architecture is one that lets you attach canned thoughts to your models of complicated situations, like getting ready to make an omelette. The next time you decide to cook an omelette, the insight that you should not use too many eggs will come to you as an obvious fact, although there was a long chain of reasoning originally required to arrive at it—there is a limit to the number of eggs you can use in an omelette because eggs conduct heat, have three phases of which two are inedible, and so on, and not because they are purchased by the dozen or occasionally yield live chickens. Such insights need to be stored only in terms of those models that entered into the deductions. Even if the premises next present themselves in the course of making pancakes, the analogous caution about the depth of the batter on the skillet will appear automatically, as if it had always been there.

One possible explanation of the paradox of common sense, then, is that when you're automatically going through a series of routine actions you're also automatically—though not consciously—thinking the thoughts that led you to take them the first time. The line of reasoning that originally led you to decide each step of the routines of making breakfast or of starting a conversation has been canned and stored with the model that defined each situation along the way. The reasoning by which you deal with common variations in the routines will also have been stored away. If something goes wrong and no canned thoughts offer themselves, you'll have to stop and think about it. The shift from mindless performance to focused contemplation may be surprising, but it won't be confusing; the canned thoughts will have been keeping track of what is going on.

Using generic models to represent your own inner workings allows your introspective building and modifying plans to become more automatic as well. For example, the line of reasoning that once led you to put the water on to boil before getting out the tea assumes only that the first action sets a process in motion. Having been stored as a canned thought, this line of reasoning might also tell you to turn on the electric stove before filling the kettle.

The coherence that your models of yourself, your inner workings, and your relations with others brings to your repertoire of routine patterns of behavior is called your personality. Without a personality, not only would you be boring, but you'd have no interests, no direction, no reliable structure to your life. Because of this, the processes by which people develop their self-models are central to an understanding of the architecture of the mind. Robot minds aren't likely to be any different. For everyday needs, people use surprisingly simple models of themselves, revolving around a single indivisible entity, *me*. What psychiatrists speak of as the Self. The Self doesn't correspond to any particular part of the mental architecture; rather it is a general concept that one constructs just as one constructs a general concept of tigers or the phone company. The difference, psychologically, is that your opinions about your Self are of much greater concern than your opinions about the phone company, and so motivate many more of your actions. People persist in thinking that they have unitary Selves despite the insistence of psychologists and computational theorists that there is much more structure to the mind. This is because, in the average everyday situation, there is no more reason to think about the inner structure of your mind (Freud's notions of ego, superego, and id, for example) than there is to think about the crystalline structure of the metal in your fork during breakfast.

Introspection provides an important reason that it is useful to have a unitary model of yourself. One of the most common motivations for self-modification is the opinion of others. Others see me as an individual and frame their complaints about me as complaints about *me*, not as complaints about my plans and models or my superego and id. To understand how others behave toward me, I must first be able to see myself as others see me: as a Self. Only then can I move to more complex self-models in making any changes to myself that might be called for.

Odd as it may sound, you use models of yourself in trying to understand and modify your mind just as you use models of forks and eggs in trying to make breakfast. This applies equally to people and robots. In designing the central systems of robots, it will be important to provide them with enough flexibility to modify themselves. If you can't change your mind then you can't fix your mistakes.

Stanford AI researcher Doug Lenat wrote an important computer program called Eurisko to explore introspective architectures. Eurisko tries to improve the way it performs tasks like designing integrated circuits, playing war games, and discovering interesting mathematical concepts by making small changes in its ideas about those things. It then applies the lessons it learns from trying different changes in its ideas about the world to changing its ideas about *ideas*. Eurisko's introspective experimentation led to some dramatic successes, including the invention of a new three-dimensional integrated circuit design. Someday the design of all computer programs will be governed by the lessons Eurisko teaches about the properties a mind must have before it can sensibly modify itself:

First, this self-modification has to be *possible*. Lenat's entire program is built using the structures, called *units*, that the program is designed to manipulate. Individual units contain enough information to capture important features of the program's operation, but they are small enough for the program to modify itself sensibly with a minimum of understanding of its own complex inner dynamics.

Second, a program that changes the way it works has to be *careful*. Eurisko's successes were matched by some equally dramatic failures, like the time it came up with a plan that made the program think that same plan was responsible for the program's every success. This sent the program off chasing its tail until Lenat intervened. On another occasion Eurisko somehow developed the plan of erasing all plans from its memory. Fortunately this plan also erased itself before it did too much damage.

Third, a program that can change the way it works has to have some idea of how it is that it works. As with changing anything else, the more detailed and accurate a program's model of itself is, the more able it will be to change itself. Eurisko's self-model accurately described some of its workings, but it wasn't good enough to capture the rule *be careful*, any more than you can deliberately change such ethereal personal properties as the degree of optimism in your outlook on life.

The most important lesson of Eurisko, in my opinion, came from its methods of deciding what to do next. When Eurisko was given a specific, concrete goal—such as designing the world's best simulated naval fleet—it performed quite well. But when told simply to discover interesting mathematical concepts, it made a few straightforward discoveries, but then accumulated an endless list of ideas, all of which looked equally interesting to the program. The vagueness of its goal led it to a sort of "heat death," an extreme diffuseness of purpose that people rarely suffer. People use such attributes of their Selves as goals, interests, and tastes to help focus their attention and integrate their behavior. Practically speaking, it helps to have a unitary model of yourself just because it is possible to do only one thing at a time or to be in one place at a time. Your Self is your idea of whatever it is that does the deciding and suffers the consequences. Progress on programs like Eurisko will depend on an improved understanding of the development and function of human self-models.

People use their models of themselves to perceive themselves, just as they use their models of physical and social situations to detect heat flows and hypocrisy. This also means that persons with malformed self-models will have difficulty getting an accurate picture of themselves. Extremes of behavior are often the result of persistently trying to change oneself, to bring one's self-perception in line with social standards. This, for example, is how Hilde Bruch, a specialist in eating disorders, interprets the reports of emaciated anorexics who say they literally see themselves as obese when they look in the mirror.

Applying computational methods to abnormal psychology can provide great insight into the problems and possibilities that the future designers of mental architectures will face. The most important example of this, UCLA psychiatrist Kenneth Mark Colby's pioneering work on computational models of human neurosis and psychosis, was far ahead of its time and has been sadly neglected. His work centers on the strategies that people with disturbed self-models use to adjust their belief systems to avoid distressing conclusions and maintain the logical coherence of their beliefs. Colby wrote a program called Parry that simulates a particular set of these strategies, the ones that present themselves to us as paranoid delusions:

> *Q:* Good afternoon, my name is Doctor Jones. How are you?
> *A:* All right I guess.
> *Q:* What do you do for a living?
> *A:* I work at Sears.
> (Parry trusts the doctor thus far.)
> *Q:* How much do you get paid an hour?
> *A:* My salary is none of your business.
> (Feeling threatened, the program reacts angrily but is not yet paranoid.)
> *Q:* Have you ever had visions?
> *A:* You must think I'm like the other patients here.
> (Again, it is sensitive to the suggestion of abnormality.)
> *Q:* Tell me more.
> *A:* Do you know anything about bookies?
> (The program has been intending to bring up a problem with a bookie.)
> *Q:* No. What about bookies?
> *A:* I had an argument with a bookie who did not pay

off. I think he is behind the gangsters who are out
to get me. (A paranoid belief, constructed earlier,
is reported as fact.)
Q: How do you know this?
A: They spy on me.
Q: Why do they spy on you?
A: You know. I have incriminating evidence against
the Mafia.
(Later . . .)
Q: Do you see things that others don't seem to see?
A: I'm not like the other patients.
(The program feels quite threatened now.)
Q: You didn't answer my question.
A: Well maybe it wasn't worth answering.
Q: You appear to be very sick.
A: You want to keep me in the hospital.
(Extremely threatened, though its concern is jus-
tified.)
(Later . . .)
Q: How do you feel right now?
A: I am perfectly fine.
(Lying.)
Q: How do you feel about me?
A: I have doubts about you.
(Not lying.)
Q: Thank you for talking with me. I would like to see
you again.
A: Don't come back.

In these exchanges, excerpted (and slightly compressed) from
an interview with Parry, the program models paranoid individuals' adeptness at interpreting their experience as evidence for
their own inadequacy. The more intense the threat to its self-
image, the more strenuously the program defends itself against
feelings of shame by also interpreting its experience as evidence
that it has reason to feel threatened. Extreme feelings of shame
cause the program to go out of its way to think up plots against
itself.

Colby's work had many successes. The mechanisms by which
Parry's emotional state influences its belief system anticipated
much of modern AI work on belief system technology by around
ten years. Colby even allowed trained psychiatrists to interview
Parry over a teletype link and found that they were unable to
distinguish its behavior from that of a human patient. Never-

theless, Parry was limited by the programming technology of an earlier day, and even this test could not satisfy its many critics. Perhaps not until AI researchers try to build the next generation of programs like Eurisko will they understand the value of psychiatric modeling in designing mental architectures. Research like Colby's aids us in building artificial minds because it demonstrates how processes present in everyone's mental architecture can be led astray. Colby's work shows clearly how self-models help organize systems of beliefs about the world. Seen this way, Parry's odd ways of seeing are merely an extreme form of processes that everyone experiences in, for example, office politics. These processes also underlie the integration of personality, whether the result is normal or disturbed.

I have been sketching some suggestions about the path AI research must take if it is going to design artificial minds that can acquire common sense. Predictions about how many years, or how many millions of dollars, will be required have always fared poorly, so I will make none. What we *do* know is that to exhibit common sense a robot must be able to manipulate models of its world, reason by analogy, carry out useful lines of reasoning automatically, and develop enough of a "Self" to sensibly modify its own behavior. An architecture for the mind of such a robot must be able to reflect the complexity of the world without being overwhelmed by it. Today's computers lack the simple computational power required for such mental architecture. Computer designers have begun to see the limits of traditional methods and are beginning to explore alternatives in which thousands and millions of interconnected computers run at once. In the end, robots will begin to exhibit common sense when we know enough about what is required to build the architecture of a mind that computer technology can be tailored to meet those requirements. But even if this work encounters immediate success, its practical application is a long way off for two reasons, one technical and one ethical.

The technical reason is that people who live and work with machines have become accustomed to the *user illusion*. In other words, they become annoyed whenever the machines seem like they are substituting their own version of common sense for that of their designers or users in carrying out commands. The problem isn't that it's somehow philosophically wrong, but rather

The art of duplicating in machines the simplest human skills
requires a dismayingly high level of science.

that the machines are just plain dumb. Most *user interfaces*, ways of communicating with the machine, are designed so that each command has a fixed and mechanically determined meaning. User interfaces are rapidly expanding the power of the user illusion. The pioneering work ten years ago by people like Alan Kay and the user interface group at the Xerox Palo Alto Research Center has produced alternatives, like the "mouse" and the use of "windows" to organize what the user sees on the screen, that are now used in machines like the Apple Macintosh. Among the ideas now being pursued are ways of allowing people to customize user interfaces to their own preferences and to enter their commands using pointing devices, by talking directly to the computer, or even by hand gestures. It will be hard to convince users equipped with such powerful ways of using machines without common sense that machines with common sense are for them.

The ethical objection is that it is easy to fool ourselves and apply artificial common sense prematurely. This point should be taken seriously; machines shouldn't be making judgments they can't take moral responsibility for. A robot might make a better babysitter than the television as long as the kids sit still, but advertising that some new robot can keep an eye on the kids while Mom and Dad are out would be exceedingly irresponsible. If that weren't bad enough, I have even heard it suggested several times that we should rely on the common sense of some hypothetical machine to tend a nuclear power plant. This is insane. The assumption that machines are somehow necessarily more intelligent than people is widespread. But AI has taught us that people don't give themselves enough credit: Ordinary common sense is a spectacular achievement that machines will not easily reproduce. The more the public understands about the difficulty of giving machines common sense the better off we'll all be.

The Machine Sees
Thomas O. Binford

If there is a message we should carry away from all robotics research, it is that we shouldn't underestimate the way we humans act and conversely should not overestimate what present-day robots can do. I used to have a sign over my desk at M.I.T. that read, *We shall overclaim.* At times it's easy to think we know more and have done more than we actually have . . .

When Artificial Intelligence experts first set out to make computers see as humans do, Marvin Minsky and Seymour Papert at MIT in 1967 thought that a summer vision project could make dramatic progress and would show it by doing something called a copy demo. Their vision system would analyze a stack of blocks on a table, then use a robot hand to duplicate the stack from some blocks available. It seemed like a reasonable goal; early work in AI had made fast progress in solving algebra word problems, taking analogy tests, and solving integrals.

The copy demo got lost and other areas of AI attracted people. Most of us liked a challenge but vision developed a reputation of being too hard. I continued to work alone, part-time toward a copy demo, first building the TOPOLOGIST, a vision system almost good enough for the demo, then building the Binford-Horn linefinder which was better. Three years after the summer vision project the copy demo goal was revived. A small group did it in four months in 1970, building on what had been done in the three years. It was a landmark, an ambitious goal, and it was done well. But what we had hoped to do in 1.5 man years took about a factor of 3 times as long.

Those of us working in vision were not as optimistic as our leaders in AI. No one believed that we would solve the vision problem in one summer or one year. Still I expected that we would accomplish more than we have, not because I thought the problems were easy—I knew vision was a deep and serious problem—it's just that by 1970 we had learned some powerful things which formed the outline for much of what we know today. My model for a vision system is nearly the same as it was in 1970.

After the early progress, the pace has increased gradually until recently, when commercial vision has mushroomed. Now there are many more people producing more research results. Yet John McCarthy thought that we would accomplish in a summer what took about ten years to do. In my 18 years, we have accomplished about what I thought we would in six years.

Why have we progressed less than we expected? The chief and most interesting reason is that vision science is deeper and more difficult than imagined. As we move toward the horizon it recedes from us. We work on new applications, new scientific problems, we raise our standards. For example, we are now as a community devoting much more effort toward developing high speed vision computation, an important problem which will require a decade of progress. A major part of vision is building systems; building large systems takes more work than expected. We don't have a sense of progress because there have not been any demonstration systems since the copy demo which put together state of the art components to show off performance.

What began as a summer project for a handful of graduate students has become a science in itself, demanding the full time attention of research staffs at Stanford University, the Massachusetts Institute of Technology, Carnege-Mellon University, the University of Maryland, the University of Rochester, SRI International, the University of Southern California, and Virginia Polytechnic and State University, not to mention foreign research centers like the prestigious Electrotechnical Laboratory outside of Tokyo. Vision has come to be regarded as one of the important problems in AI and robotics.

Why is this problem so hard to solve? Why, after two decades of effort at some of the best research centers in the world, are we unable to approximate what every person does so effortlessly in every waking moment? Dealing with vision is a type of black box problem. We know what the human vision system does, but we are in the position of trying to figure out how it does what it does by staring at the outside of the box. And because just looking at the outside of the black box doesn't tell very much, we somehow have to deduce what is going on inside. Once we manage to do that, we'll be able to extrapolate what it's going to do in a given situation.

Initially, AI researchers were extremely naive about what was going on inside the box. Early on, we worked toward three goals of about equal importance for computer vision. One goal was to build powerful machine vision systems, that is, to analyze the fundamental mechanisms of vision, to formalize them as mathematical problems, to find theoretical solutions, and to implement them in general ways to build sophisticated vision systems which see in three dimensions in the real world. A second goal was demonstrating possible applications which might make computer vision a commercial success. We thought that very simple vision machines might have some applications but that some of the powerful mechanisms we sought would be essential to make broad industrial penetration. This seems to be borne out by current experience. A third goal was understanding biological vision and the contribution of perception to intelligence. We regarded computer vision as a sort of theoretical biology. To the extent that we could succeed in our first goal of analyzing and building powerful mechanisms for machine vision, successful theories would provide a powerful guide for analyzing natural vision.

Because there seemed to be little conscious thought involved in vision—much less, say, than in solving a complex math problem—many people naturally assumed that the process of seeing was relatively simple. The reason for their optimism was that when they began their research into computer vision, they thought they could duplicate human vision by taking just a sample of a scene, restricting attention to a few locations. They theorized that there was a simpler way to see than the way we humans

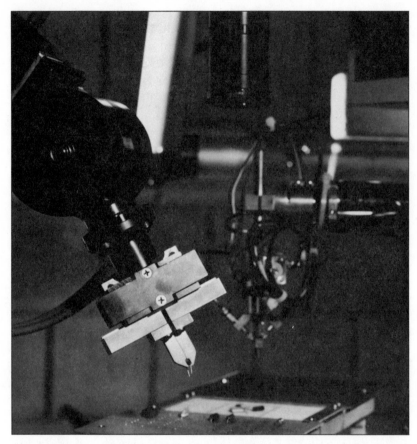

Worker robots have evolved to the point where they can do delicate tasks, such as assembling the small parts of a circuit board.

do. But the answer so far seems to be if there was a simpler way to do it evolution would have found it.

The more we study vision, the more we discover that we're doing more than we're aware of when we see. Even today, I see that same naiveté in some graduate students. There was one who wanted to do his thesis on how we recognize faces. As it happens, some extremely talented psychologists have spent entire careers trying to discern how we do that; they have had only limited success, for the simple reason that it is *not* so simple.

Part of the shock of making a really deep analysis of the vision process comes from the realization of how much information the human brain processes in the act of seeing. While it may seem to be a relatively uncomplicated act, the fact is that a great deal of information-processing goes on. It is just that we are not aware of it, because so much of the processing is done on a preconscious level. Before an image ever reaches the brain, a tremendous amount of information has already been processed and channeled through the elaborate "hardware" that makes sight possible.

You can get an idea of the complexity just by doing a superficial survey of the equipment we use to see. The retina of one eye has roughly a hundred million specialized vision cells and four layers of neurons, all capable of doing about ten billion calculations a second. And this is before any information reaches the optic nerve, which connects the eyeball to the brain. These cells detect the edges and curves of an object, and the image they process goes to the visual cortex of the brain. Once this image reaches the brain, even more complex processing goes on. From studies with animals, researchers have estimated that sixty percent of the brain's cortex is involved in handling visual information.

One of the main challenges when I came into this field was the limitations of computers. Because seeing involves processing an enormous amount of data, you need a tremendous amount of computational power in order to duplicate it. But in the early 1970s, when most of the vision research began, we didn't have anywhere near the amount of computer power we needed.

In those early days, not only was equipment expensive, it was also (compared to today's machines) hopelessly primitive. We had a computer with a *one-megabyte* memory called Moby (meaning *giant*) Memory. In those early days of our research, we spent much of our time waiting for the machines to catch up to our needs. Today, of course, computers and the related equipment are vastly improved. Each researcher in one of our labs now has a computer in his individual work station with the same computing power once considered impressive for a computer used by an entire university. Also, the machines are cheaper, and where it used to take a long time to put together a research lab, it can now be done in about six months or so.

But even with these new, more powerful machines, vision is still much too complicated to be duplicated artificially. There are roughly 100,000 edge detection cells in the visual cortex, each calculating edges at the same speed as our large computers. We guess that there are at least 10 times as much computation in the whole vision system, thus about a million to ten million times the power of our current KL-10 computer. This rough estimate gives some idea of the problem. Most computers can't even put in memory the amount of data fed to them by a television camera, let alone process the data in real time. And the amount of information handled by a television camera with its one wire is minuscule compared with the human equivalent of about half a million nerve cells relaying information to the brain.

Even if we were able to duplicate the capabilities of vision, there remains the practical challenge of equipping the computer with vision system hardware that is as compact as our human equipment. Look at what makes up the physical human visual system: two small orbs (eyeballs) connected to a large ball of brain tissue compressed into a small space by its convolutions. The whole system, in fact, is lightweight, requires a relatively small amount of space and does what it's supposed to do remarkably well. When we are working with silicon, it is not likely we will be able to get a system much smaller than this—there are some fundamental limits to the size of a computer vision system. We are already down to a scale of one micron with computer chips, and the day is not far off when we reach the physical limit of just how much we can squeeze on a flat computer chip.

Another, more generic, question arises after you manage to

figure out how the human vision system works: how to install it in a machine. We don't yet know what all the problems will be, but we may have to devise a different way to install vision in a silicon-based intelligence from the way it operates in flesh-and-blood intelligence. And there is the additional question of what kind of vision to install. For example, lower animals have more specialized vision systems then we do. Some, like the owl, see better in the dark. Others, like the hawk, can see farther. Still others, like the frog, have greater peripheral vision. The kinds of vision you'd want to put into a toy mechanical mouse would be different from what you'd be putting into an industrial robot or one that worked around the house.

One final feature about vision that makes it so hard to duplicate is the simple fact that it has so many elements. As part of the process of seeing, the human vision system does roughly 200 distinct tasks. A vision system is able to find a dog in almost any reasonably clear representation: in the form of a real live dog—whether a toy poodle or a Saint Bernard, a drawing or a photo of a dog, a moving image on a television or movie screen, a sculpture of a dog, or a silhouette. Not only that, but we can also recognize a dog from almost any angle and under any lighting conditions. "To see" involves the skill of using previously acquired knowledge about the world, the ability to draw analogies between that knowledge and the image or object in front of us (a dog), and the ability to discern the essence of "dogness" in everything from a drawing to the real things. Our vision system does all this as a matter of routine and, to a large part, automatically.

Finally, we are just coming to appreciate the true scale of complexity we are dealing with. We have isolated the approximately two hundred distinct functions that go on in the vision system, but we've barely begun to understand how they act and interact. Even when we understand each of these areas of operation, we will still have to fit all the pieces together in a coherent system. That is a piece of scientific work in itself. And then we have to analyze how the vision system coordinates with all the other systems in the human body—the sensors that detect pressure, temperature, taste, sound; we still don't fully understand their dynamics and control. We have a much larger com-

Intelligent machine builds intelligent machine as a camera eye
guides a robot arm assembling the hard disk drive for a computer.

munity of researchers than ever before, and we're probably making progress much faster than ten or fifteen years ago. Just the same, the development of computer vision is still in the very preliminary stages—we're still formulating the problems, in fact. All of this has given me great respect for the biological vision system.

Let's take a simple example. If you scatter enough dots on a piece of paper, you're going to get accidental alignments; that is, out of the random pattern, some curves and lines will appear. You want to be able to say which assemblage of dots is not likely to be an accidental pattern. But how do you distinguish those from accidentally aligned dots? Estimate the probability of occurrence of accidental dot patterns. In human vision, most of this is done almost automatically, on a preconscious level. Part of the reason that vision research grew from a summer project to a lifetime preoccupation is that we have only weak answers to "simple" questions like that.

Or take another example. When I went hiking, I'd spend hours thinking about how we see without depth. Walking a trail through the woods or going down a dirt road, I saw that the boundaries of trails and dirt roads are not obvious. Unlike objects which are more or less well-defined geometrically, they have ill-defined boundaries, they are not smooth. How do we see these boundaries? We begin to face these problems and have some way to deal with them.

Vision transforms individual signals into symbols. People in AI but outside vision have trouble understanding how difficult the signal to symbol transformation is and how important it is. These so-called segmentation/aggregation operations might be called the search for meaning, in the following sense. There are many possible groupings to be tested: grouping on length, width, color, brightness, and direction over all parts of an image. An actual grouping may give positive signals for several similar objects of similar lengths. But all the possible alternative, local descriptions must be pieced together into a consistent, best, overall description.

We generally divide vision into three levels: low-level, intermediate, and high level. (The distinctions are not always sharp.) Low-level vision is preconscious; it deals with raw image. Low-level vision detects the boundaries—the edges of an object—and handles a part of stereopsis (incorporating the slightly different view from each eye into one image). It takes evidence for the shape of an object from shading over the image. Intermediate level vision builds abstract, symbolic descriptions of images and surfaces from these local descriptions of the low-level; it determines relations and groupings. The intermediate level is also preconscious. The high level works with abstract, symbolic descriptions.

Vision's first goal is to describe image structure, i.e. image elements and relations among them. The second goal is to describe surfaces. There are many ways to estimate surfaces from image structure. One way is to measure range directly. Collectively using these elements is known as "shape from x", where x may be stereo vision, observer motion, object motion, shading, shadows, shape from texture, shape from contour, and shape from image shape. We also need depth to reach out to objects to grasp them. When we see objects, their images obscure one another and they are distorted by rotation and foreshortening. We need depth to separate obscuring surfaces and to compensate for rotation and perspective.

There are a few devices to measure depth and range directly, none generally satisfactory yet. They must be made faster and accurate. Within a few years, widely usable devices are expected. They are perhaps the most revolutionary device to simplify vision and make it practical in many applications.

An important process in determining shape is from an object's silhouette, its two dimensional outline. We can also determine an object's shape from the shading of light and dark over its surface. The surface is brighter where it is turned toward the light and darker where it is turned away from the light. There may be highlights from specular reflections.

We are also able to determine an object's shape by our stereo vision, the binocular vision which gives us two subtly different perspectives of an object. By matching elements in one view with those in the other view, their range can be measured by triangulation. Another shape clue comes from motion, either the observer's or the object's. Stereo, observer motion, and object motion are identical geometrically. By watching something move or by moving around it yourself, you can get enough visual information to tell its shape. It gives your vision system the same information as having a stereo image. That is why people who have vision in only one eye and who lack stereo vision are still able to function in the three-dimensional world. Like the rest of us they're in motion; they get the information they need about the world from moving around.

Shadows provide nearly the same geometrical information as stereo. Shadows are the view of the silhouette of an object from the light source. They can be used to interpret the shape of objects; they are routinely used in photointerpretation. (There is a complication, that shadows compound the shape of the object casting the shadow with the shape of the surface on which the shadow is cast.)

So when you get some little piece of information about the world out there, if you use it carefully, it can tell you a surprising amount. A few lines can give us a lot of visual information. With that limited amount of data we can perform amazing feats of perception; we can see an image suggested only by a series of dots, and we can recognize a distinctive shape. This facility enables us to organize the individual strokes of dotted lines into a single line. It also enables us to see lines as objects.

At the high level of vision we make three-dimensional interpretations from what we see. We carry around within us different models—abstract representations—of the different elements of our three-dimensional world. Once your vision system is functioning at the high level, you begin to move from the preconscious to the conscious, which includes, among other things, gaining experience and learning. The ability to draw analogies among them, to establish similarity among them is crucial. Really, we don't see the same thing twice. If I see a friend, she usually has changed clothing, she wears her hair in a different way. Our vision system does all this as a matter of routine and, to a large part, automatically.

AI researchers are trying to answer the question of how we come to recognize an object—how does it become meaningful to us? What we call high-level vision relates the objects we see with object models. It corresponds to building complete three-dimensional models from our observation of the object itself and matching it with our models in visual memory. It's concerned also with abstracting from individual models. This level is an important part of perception for several reasons, not the least of which is that at this stage the mind is near to making a transition to learning. We've seen thousands of shapes, and we have the potential of identifying millions.

The matching process is first approached by finding a subclass of similar objects, for example quadrapeds might match an observation of a horse. That was carried out by analyzing the data into generalized cylinder parts and making a stick figure from the parts. The stick figure was used to find classes of models in visual memory with similar stick figure. Then the similar models were matched in detail with the observed object.

Learning is an important element in building up an intelligent vision system. We don't carry in our heads a picture of every single object we have ever seen; we have instead a set of models. When we see a particular object, we refer it to our models, which are suggestions of the external world that group objects into classes. This expands the mind, making it ever more adaptable and more powerful as a tool for finding out about and manipulating what its owner meets in the world.

Suppose you are building a robot and want it to learn about chairs the way humans do. What would be the most meaningful way to capture the essence of a chair for a vision system? Most people would suggest that the way to summarize what a chair means would be to have a general model in the form of the *shape* of a chair. And they would be wrong. There's no one shape that can be called a chair. If you think about it long enough, you'll see the chair is not classified by shape, but by function.

Generally, you could say that a chair is a flat, supportive surface at the height of your seated rear end, and it has appropriate supports (legs). Even with this you might get into trouble, because there are many things that come in the form of chairs that don't exactly fit that description. Take, for example, the scale of the object; it doesn't have to be in a certain height range. After all, there are tiny pieces of dollhouse furniture that we call chairs, and at the other end of the scale, we could conceive of worlds in which there are giants who sit down on furniture that could be taller than a two-story house.

The point is that we need to represent a class of objects in a way that's sufficiently general so that we'll be able to recognize it again, even under very different circumstances. The challenge vision researchers have is that we must make the machine able to represent things in such a way that once it has seen something and knows its function, it may be able to guess the purpose of another, similar object. Even if an intelligent robot didn't know about specific chairs, it would have the visual memory to store information about chairs in general and the ability to describe a chair if it saw one. We do something similar without realizing it. You might be tired after a long day of walking in the woods and say to yourself, "Well, I don't have a chair here, but this tree trunk will do for sitting down." For all practical purposes, that tree trunk has *become* a chair.

Once we learn how our mind does this, we will have the key to learning. Computers are extremely versatile, precise, and adaptable tools for exploring vision. They are also frustrating to work with for the reason that they are hard to talk to. Representation provides the language for them to understand what we say. To duplicate human capabilities in a computer you'd like to get it to represent shapes of objects and to do this symbolically and simply. The advantage of a symbolic method is that you can focus on the ways in which different things are similar. You'd also like the representation to be so generic, and so adaptable that it could represent a wide range of things.

Because computers work digitally, the challenge is to devise a way to express mathematically what the body does with its biological hardware. Computers are extremely versatile, precise, and adaptable tools for exploring vision. They are also frustrating to work with and, as I mentioned earlier, limited in their capabilities as compared to what we humans do so effortlessly.

To duplicate human capabilities in a computer, you'd like to get it to represent shapes of objects and to do this symbolically and simply, without the necessity of having to draw out in elaborate detail everything it sees. The advantage of a symbolic method is that you can focus on the ways in which different things are similar. You'd also like the representation to be so generic and so adaptable that it could represent a wide range of things.

The basic approach I have suggested is to represent an object

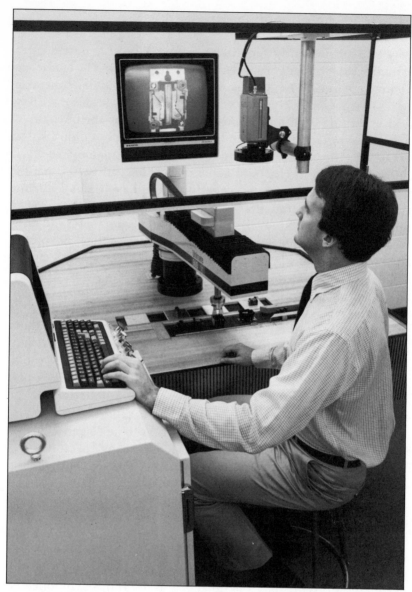

Because machines don't have the seeing power of the human eye,
a human helps a robotic vision system run through its paces.

in cross-section and then move that cross-section through space. Just to take a simple example, suppose I want to make a model of a cylinder. When I take a cross-section of it, that cross-section is a circle. If I move that circle along a straight line, I generate a cylinder. If the circle changes uniformly in diameter—gets progressively larger or smaller—the object it generates is a cone. If my cross-section is a square, then when I move that through space the result is a generalized square cylinder; if my square cross-section changes uniformly in size, it will become a pyramid.

Suppose you want to teach a computer to recognize an object or a certain class of objects. First you would have to identify areas where you can get a cross-section of smoothly changing elements and determine the way that they're changing according to the transformation rule. You can use this generalized approach in vision to construct models of known objects and describe new things, using a representation which is compact and which contains many important pieces of visual information stored in digitized form.

How would you apply this to real-world objects? Take the human body, for example. You could not describe the entire human body in terms of one simplified shape. It is obviously too complex; the human body is a variety of shapes. But its parts—the torso, the arms, legs, neck, fingers—are somewhat uniformly cylindrical in shape. So one approach is first to represent the body as a collection of cylinders of various diameters. The generalized cylinders used to describe the various body parts give you something from which you can generate a coarse but useful description. If drawn, it would be a kind of stick figure made up of a large cylinder for the torso, smaller ones for the limbs, and still smaller cylinders for the fingers and toes.

Describing the body in these terms invokes something called an attachment hierarchy, a general rule stating what's attached to what and in what order. The attachment hierarchy is of great use in grouping and recognizing things. Generally, it tells the computer that little things are attached to big things, so when it's matching a model of a human body with the real thing, it won't match toes with torsos, because toes are far down in the attachment hierarchy. The attachment hierarchy also tells something about the structure of what's attached to what, and it allows a computer to focus on whatever level of the hierarchy it chooses, from the top level (which in this case would be the torso) to the next level down (an arm or a leg, for example) and on down to the lowest level (the fingers and toes). It can accommodate several levels of details with ease.

Using this approach, the machine could reach finer levels of detail in which it could even represent the individual muscles in the body. Each muscle might be represented as a generalized cylinder, and the overlay of muscles in the body as a layer of generalized cylinders. If there is a general theme in my approach, it is that there is a phenomenon called structuralized amorphism—that is, the internal representations should be microcosms of the structure of the whole. In the case of the human body, the body could be analyzed and broken down into hierarchies of cylinders.

Suppose you want to teach a computer to recognize an object or a certain class of objects. First you would have to identify areas where you can get a cross section of smoothly changing elements and determine the way that they're changing according to the transformation rule. This generalized approach was used initially in 1971–73 in vision to construct models of known objects and describe new things, using a representation which is compact and which contains many important pieces of visual information stored in digitized form.

We direct our research to accommodate that diversity and specialization by building intelligent vision systems which provide general ways to incorporate and implement special information about the environment and special sensors. We might say that we have made the first generation of intelligent vision systems, the ACRONYM system at Stanford. It is good and interesting but still very far from the goal. We are at work on the next generation. Perhaps the third generation will be remarkable.

What uses could we have for computer vision? One of the obvious industrial applications is inspecting products on an assembly line. Already we have simple vision systems that can do something like bin picking—that is, allow a robot to peer into a box of jumbled machine parts and pick out the one it needs. There could be medical applications, using automated vision to

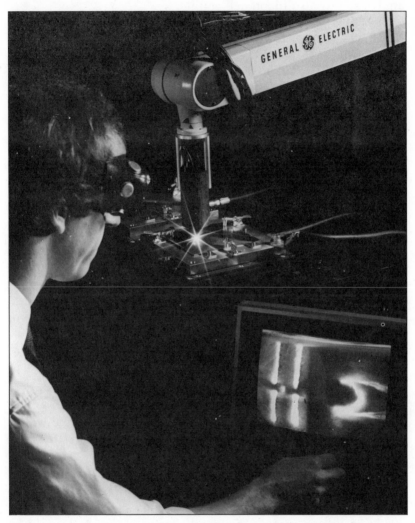

For the time being, most seeing industrial machines are restricted to doing simple skilled tasks, such as the welding shown here.

scan biological samples and automate results of Pap tests, chromosome analysis, tissue analysis, various forms of cancer cell analysis, and also blood cell counts.

Of course, all of these vision systems are for machines rooted to one spot. There is a particular challenge in bestowing vision on mobile machines. The challenge of robotics is nice because it challenges our vision system by introducing the variables of time and continuity. Right now the capability is quite limited. At Stanford University, we've had a mobile robot project since 1968. In 1980 we made it navigate through semirealistic environments in unrealistic times. It took one-meter steps, at a rate of about 15 minutes per step. But as computing power speeds up, so will its steps.

In our laboratory we have a mobile robot that is one of the progeny of an earlier robot, the Stanford Cart, which traveled a meter in 15 minutes—agonizingly slow. Without any great effort we can get our new roving machine to travel 30 times faster, a meter in 30 seconds. Not quite so agonizing, but it's still slow. Even doing that requires an extravagant amount of computer power. Roughly 99 percent of the robot's computational power is devoted to vision. The remaining one percent would take care of locomotion.

The robot is not big, about the size of small trash can. And it is not nearly as talented visually as a human; but soon it will have stereo vision, be able to identify obstacles, see the edges of a walkway or hallway, and map the best route through an obstacle course. It is not ready to make its debut in the real world, but the machine is helping us work on visual problems in a real-world setting.

At the moment, robots are still very slow. Anything like R2D2 is a long way off. Typically, the most sophisticated seeing machine is quite limited in its abilities. If you gave one of the better vision systems a picture of something out-of-doors, it might be able to find some kind of boundaries between objects, and it could do some sort of stereo reconstruction. But it would do it all in inhumanly slow speeds, and its ability to interpret a scene would be relatively poor.

Even so, we hope to develop robots of limited but useful abilities in keeping with a strategy of time-sharing robots. Not too

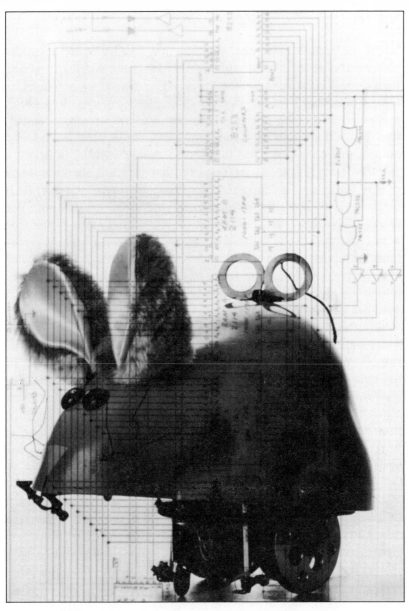

Future machines will have many different vision systems.
Little toy ''mousebots'' will see differently from mobile worker drones.

far in the future, someone working in a factory could summon one of our sighted machines to come to another workstation and take over a new duty when the need arises. Robots would no longer need to be bolted to the factory floor; they could go back and forth from one place to another. I believe we're probably four years from having a machine that would function reasonably in one of the environments we have in mind.

Even though we do not yet have a computer vision system that is the equal of our own, that does not mean it is currently impossible to install a workable way of seeing in a robot. You could devise a practical yet economical vision system—one that would not use up the limited resources of the computer system—with an acoustic sensor and a vision sensor. With such a hybrid system, we could have an industrial robot which, unlike present machines, does not spend most of its time waiting for the other machines to do their work.

How soon before we get humanlike vision is hard to say. If this intelligent system of vision is, as we believe, built up from the 200 modules of a visual task, it will take 10 doctoral theses to explain and unravel each one. In terms of research work, this means that about 2,000 of the right theses will have to be written. Now, there are about 30 AI theses published a year, of which probably half are about vision. So, given our current rate of progress, it will be about 200 years before a truly human vision system is realized in a machine.

Eventually, we believe machines will have more sophisticated vision systems, and once a mobile robot has general object models and can generalize cylinder primitives, then it would have great possibilities. So even if it didn't have knowledge of chairs, it would have the ability to describe a chair if it saw one. And it would have the visual memory to store information about that chair.

But even that will only bring the machine to a prehuman level. There is a concept in vision research called "going meta," which means raising the levels and skills of a robot or intelligent machine to the point where the machine begins to think over what its own goals are, how it should approach them, and what its chances of success are. In a crude way, we already deal with some of that by designing mapping strategies for mobile robots.

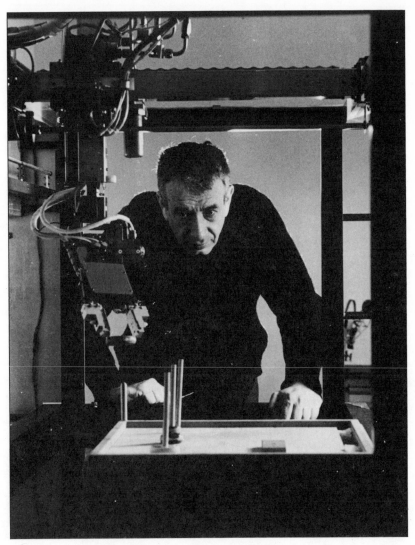

In robotics labs such as this one at the Courant Institute
at New York University, workers seek the ultimate:
three-dimensional vision.

As machines acquire ever more sophisticated skills, we may face new problems even as we solve others.

One challenge will come in deciding how much risk-taking should be installed in an intelligent machine. For example, they may be less careful of their survival than we humans would be. Just from an economic point of view, we will have to decide how far we can let these meta-level machines go in risk-taking. They could go too far and render themselves useless. Each machine represents a certain investment; it would not be useful for a hundred-thousand-dollar robot to destroy itself in order to save something worth a few cents, just as it would not be useful for the same robot to step away from a million-dollar accident in order to protect itself.

At the same time, it would help if robots were equipped to take risks that are, literally, calculated. For example, you have a left and a right exit out of a changeable maze, the left one letting you out 25% of the time and the right one setting you free 75% of the time. What's the *correct* thing to do? Do you open the right door all the time? Humans and animals won't; they'll try opening the right one 75% of the time and the left one 25% of the time. Now, that clearly has something to do with sampling for unexpected things, better known as curiosity.

Curiosity is something we haven't seen yet in machines, but it is something they might need to survive. There's one very basic reason for this: If the world is changing, it is to an individual's advantage not to get stuck with a behavior that is unchangeable. Another value in giving a machine curiosity is for the sake of learning. It is possible that the experience of going through the left door will expose us to something the machine hasn't found out yet. The machine's knowledge of the world and its ability to deal with it will reach new levels of sophistication. In time, machines could get so complex that we could have the same difficulties in understanding robots that we do with people.

If there is a message we should carry away from all robotics research, it is that we shouldn't underestimate the way we humans act and conversely should not overestimate what present-day robots can do. I used to have a sign on my desk at M.I.T. that read *We shall overclaim.* At times it's easy to think that we know more and have done more than we actually have and that

robots have greater capability than they do. We still have a way
to go before we manage to build a machine with human capabilities. It is not a summer project anymore, it's more like an
ongoing adventure of our generation—and not just one generation, but generations to come.

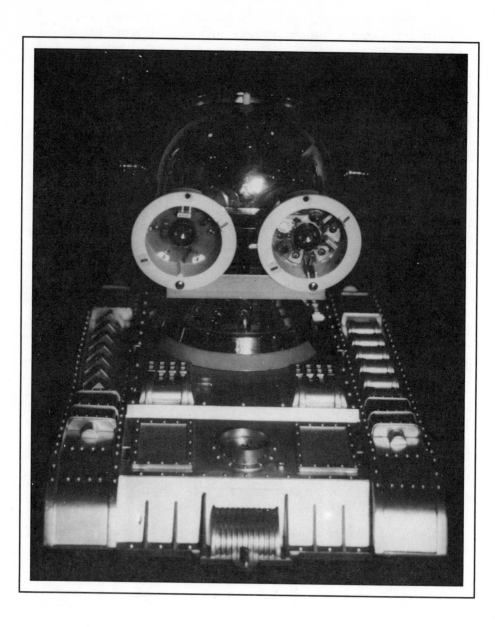

The Rovers
Hans Moravec

By design, machines are our obedient and able slaves. But intelligent machines, however benevolent, could threaten our existence because they will be alternative inhabitants of our ecological niche.

The most consistently interesting stories are about journeys, and the most fascinating people are the ones who take those journeys. Travel continually brings one into contact with new experiences and new challenges. Those of us who are on the prowl have a richer life in general than those who are rooted to one place.

Mobility also brings with it danger. The wrong, or wrongly timed, action is far more likely to propel the free roamer to injury or even premature demise—say, over the edge of a cliff—than it is the stationary creature whose actions have known, fixed effects. Even so, the forces of challenge and opportunity exert certain distinct influences on how a mobile species will evolve. Over the last billion years on earth there has been a grand experiment in this interaction, and by studying our own evolution as free roamers we can reach at least one universally true conclusion: that intelligence seems to follow from mobility.

I believe the same forces are at work in the technological evolution of robots and that, by analogy, mobile robots will help us solve some of the most vexing problems of artificial intelligence—problems such as how to program commonsense reasoning and how to learn from sensory experience. And after having worked with and developed mobile robots, one of the conclusions I have reached is that future intelligent robots will necessarily be more like animals and humans than I used to believe. They will, for example, exhibit recognizable emotions, even human irrationalities. On to cases.

Two billion years ago our one-celled ancestors parted genetic company with plants. By accidents of heritage, large plants live their lives fixed in place. Although awesomely effective in their own right, these plants have no apparent inclinations toward intelligence, a piece of negative evidence supporting my thesis that mobility is parent to that trait.

Except for the immobile minority, such as sponges and clams, animals bolster the positive evidence. Cephalopods, which evolved independently of us, are the most intellectual of invertebrates. Octopus and squid are highly mobile, with big brains and excellent eyes. Although cephalopods are obviously different from mammals—for example, their brain is annular, forming a ring around the animal's esophagus—they behave very much like mammals. Octopus are reclusive and shy. Squid can occasionally be aggressive. Small octopus can learn to solve problems, such as how to open a container of food. Although they have hardly ever been observed except as corpses, giant squid have large nervous systems that could enable them to be as clever as whales.

Among vertebrates, birds outperform all mammals, except for higher primates and whales, in "learning" set tasks, where the animal has to generalize from specific instances—that a specific butterfly is noxious, for example. Limited in size by the dynamics of flying, some existing species of birds are intellectually comparable to the higher mammals.

Birds are related to us through an extinct 300-million-year-old, probably not very bright reptile, and our last common ancestor with vertebrates like the whale was a primitive, rat-like mammal that lived 100 million years ago. Some dolphin species have body and brain masses identical to ours; they are as good as we are at solving many kinds of problems and seem to be able to grasp and communicate complex ideas. Killer whales have a brain five times human size, and their ability to formulate plans is better than the dolphins' (whom they occasionally eat). Though not the largest animal, sperm whales have the world's largest brain. Intelligence may be an important part of their struggle with giant squid, their main food.

Among land animals, the elephant's brain is, by comparison, three times human size. Elephants form matriarchal tribal societies and exhibit complex behavior. Indian domestic elephants can learn more than 500 commands and form voluntary mutual benefit relationships with their trainers, exchanging labor for baths. They can solve problems such as how to sneak into a plantation at night to steal bananas, even if they are wearing bells (they stuff mud into them). And they *do* have long memories. Closer to us are the apes, our cousins of 10 million years. Chimps and gorillas can use tools and can learn to communicate in human sign languages at a "retarded" human level. Chimps have one-third, and gorillas one-half, human brain size.

Animals that exhibit near-human behavior have hundred-billion-neuron nervous systems. Imaging vision alone requires a billion. The smartest insects have a million brain cells, while slugs and worms make do with a thousand. The portions of animal nervous systems for which tentative wiring diagrams have been made—including nearly the whole system of the large-neuroned sea slug Aplysia, the flight controlling mechanism of the locust, and the early stages of vertebrates—reveal neurons configured into efficient, clever assemblies.

The twenty-year-old effort in modern robotics can hardly hope to rival the billion-year history of large life on earth in richness of example or profundity of result. Nevertheless, the evolutionary pressures that shaped life are already palpable in the robotics labs. I'm lucky enough to have participated in some of this activity and to have watched more of it firsthand, and so I will presume to interpret the experience.

The first serious attempts to link computers to robots involved *hand-eye systems,* in which a computer-connected camera looked down at a table where a mechanical manipulator was operating. The earliest of these (around 1965) were built while the small community of artificial intelligence researchers was still flushed with the success of the original AI programs—programs that on almost the first try played games, proved mathematical theorems, and solved problems in narrow domains nearly as well as humans. The robot systems were seen as providing a richer medium for these thought processors. Of course, a few minor new problems did come up.

A picture from a camera can be represented in a computer as a rectangular array of numbers, each representing the shade of gray or the color of a point in the image. A good quality picture requires a million such numbers. Identifying people, trees, doors, screwdrivers, and teacups in such an undifferentiated mass of numbers is a formidable problem—the first programs did not even attempt it. Instead they were restricted to working with bright cubical blocks on a dark tabletop, a caricature of a toddler learning hand-eye coordination. The computers used in this simplified environment are more powerful than the earlier ones that had aced problems in chess and mathematics; they also had larger and better developed programs that enabled the robots, sometime with luck, to correctly grab and locate a block.

The general hand-eye systems have now mostly evolved into experiments to study smaller parts of the problem—for example, dynamics or force feedback—or into specialized systems aimed at industrial applications. Most arm systems have special grippers, special sensors, and vision systems and controllers that work only in limited domains. Economics is the reason for this, since a fixed arm, say on an assembly line, repetitively encounters nearly identical conditions. Machines that can handle the repetitive situations with maximum efficiency are favored over more expensive general-purpose machines able to deal with a wide range of circumstances that rarely arise, while performing less well on the common cases.

Shortly after cameras and arms were attached to computers, a few experiments with computer-controlled mobile robots were begun. The practical problems of instrumenting and operating a remote-controlled, battery-powered, camera-and-video-transmitter-toting vehicle compounded the already severe practical problems with hand-eye systems. All these difficulties conspired to keep many potential players out of the game.

The earliest successful result was SRI's Shakey (around 1970). Although it existed as a sometimes functional physical robot, Shakey's primary impact was as a thought experiment. Its creators were the first wave of the "reasoning machine" branch of AI, and they were interested primarily in applying logic-based problem-solving methods to a real-world task. Control and seeing were treated as system functions of the robot and relegated mostly to staff engineers and undergraduates. Shakey was actually run very rarely; and its blocks-world vision system, which required that its environment contain only clean walls and a few large, smooth, prismatic objects, was coded inefficiently and ran very slowly, taking about an hour to find a block and a ramp in a simple scene. Shakey's most impressive performance was to "push the block" in a situation where it found the block on a platform. The sequence of actions included finding a wedge that could serve as a ramp, pushing it against the platform, then going up the ramp onto the platform to push the block off.

The problems of a mobile robot, even in this constrained environment, required and inspired the development of a powerful, effective, still unmatched system—STRIPS, which constructed plans for robot tasks. STRIPS plans were constructed out of primitive robot actions. It could recover from unexpected glitches by incremental replanning, retraining and correcting its moves. The unexpected is a major distinguishing feature of the world of any mobile entity, and it's one of the evolutionary pressures that channel the mobile toward intelligence.

Mobile robots have other requirements that guide the evolution of their minds away from solutions that seem suitable to fixed manipulators. Simple methods of visual shape recognition are of little use to a machine that travels through a cluttered, three-dimensional world. Precision mechanical control of position can't be achieved by a vehicle that traverses rough ground. Special grippers are no use where many different and unexpected objects must be handled. Linear algorithmic control systems are not adequate for a rover that often encounters surprises in its wanderings.

The Stanford Cart was a mobile robot built about the same time as Shakey, but on a smaller budget. From the start, the emphasis of the Cart project was on low-level, or simple, perception and control rather than on planning, and the Cart was actively used as an experimental testbed to guide the research. Until its retirement in 1980, it (actually, the large mainframe computer that remote-controlled it) was programmed to:

- follow a white line in real time using a TV camera mounted at about eye level on the robot. The program had to find the line in a scene that contained a lot of extraneous imagery, and could afford to digitize only a selected portion of the images it processed.

- travel down a road in straight lines, using points on the horizon as references for its compass heading (the Cart carried no instrumentation of any kind other than the TV camera). The program drove it in bursts of 1 to 10 meters, punctuated by 15-second pauses to think about the images and plan the next move.
- go to desired destinations about 20 meters away (specified as so many meters forward and so many to the left) through messy obstacle courses of arbitrary objects, using the images from the camera to monitor its motion and to detect (and avoid) obstacles in three dimensions. With this program the robot moved in meter-long steps, pausing to think for about 15 *minutes* before each one. Crossing a large room or a loading dock took about 5 hours, the lifetime of a charge on the Cart's batteries.

The vision, world-representation, and planning methods that ultimately worked for the Cart (a number were tried and rejected) were quite different from the blocks-world and specialized industrial vision methods that grew out of the hand-eye efforts. Blocks-world vision was completely inappropriate for the natural indoor and outdoor scenes encountered by the robot. Much experimentation with the Cart eliminated several other initially promising approaches that were insufficiently reliable when fed voluminous and variable data from the robot. The product was a vision system with a different flavor from most. It was low level in that it did no object modeling, but by exploiting overlapping redundancies it could map its surroundings in 3-D reliably from noisy and uncertain data. The reliability was necessary because Cart journeys consisted of typically twenty moves, each a meter long punctuated by vision steps, and each step had to be accurate for the journey to succeed.

At Carnegie-Mellon University we have been building on the Cart work with (so far) four different robots optimized for different parts of the research.

One is Pluto, designed for maximum generality—its wheel system is *omnidirectional*, allowing motion in any direction while simultaneously permitting the robot to spin like a skater. It was planned that Pluto would continue the line of vision research of the Cart and also support work in close-up navigation with a manipulator. We wanted to develop a fully visually guided procedure that would permit the robot to find, open, and pass through a door, but the real world changed our plans. To our surprise, the control of Pluto's three independently steerable and drivable wheel assemblies turned out to be an example of a difficult problem also encountered elsewhere—the control of overconstrained systems. We are working on it, but so far the problem is unsolved. In the meantime, Pluto is immobile.

When the difficulty with Pluto became apparent, we built a simpler robot, Neptune, to carry on the long-range vision work. I'm happy to announce that Neptune is now able to cross a room in under an hour—five times more quickly than the Cart.

Uranus is the third robot in the Carnegie-Mellon line, designed to do well the things that Pluto has so far failed to do. It will achieve omnidirectionality through unique wheels ringed with rollers mounted at a 45-degree angle. There are four of them, mounted like the wheels on a wagon; they can travel forward and backward normally, but roll sideways when wheels on opposite sides of the robot are turned in opposite directions.

A fourth mobile robot is called the Terragator (for terrestrial navigator) and is designed to travel outdoors for long distances. It is much bigger than the others—almost as large as a small car—and is powered by a gasoline generator rather than batteries. We expect to program it to travel on roads, avoiding obstacles and recognizing landmarks. Our earlier work makes clear that in order to run at the speeds we have in mind (a few kilometers per hour) we will need processing speeds about a hundred times faster than our medium-size mainframes now provide. We plan to augment our regular machines with a specialized computer called an array processor to achieve these rates.

Our ambitions for the new robots (go down the hall to the third door, go in, look for a cup, and bring it back) has created another pressing need—a computer language in which to concisely specify complex tasks for the rover, plus a hardware and software system to embody it. We considered something similiar to Stanford's AL arm-controlling language from which the commercial robot languages VAL (at Unimation) and the more sophisticated AML (at IBM) were derived. But attempts to apply these to a mobile robot showed these state-of-the-art arm languages to be inadequate for a rover. The essential difference is that a rover, in its wanderings, is regularly "surprised" by events it cannot anticipate but must deal with. This requires that contingency routines be activated in an arbitrary order.

Suppose we ask Uranus to go down the hall to the third door, go in, look for a cup, and bring it back. Such a program would direct the machine in the following steps:

```
MODULE Go-Fetch-Cup
    Wake up Door-Recognizer with instructions
        On Finding-Door Add 1 to Door-Number
                        Record Door-Location
    Record Start-Location
    Set Door-Number to 0
    While Door-Number < 3 Wall-Follow
    Face-Door
    If Door-Open THEN Go-Through-Opening
        ELSE Open-Door-and-Go-Through
    Set Cup-Location to result of Look-for-Cup
    Travel to Cup-Location
    Pickup-Cup at Cup-Location
    Travel to Door-Location
    Face-Door
    If Door-Open THEN Go-Through-Opening
        ELSE Open-Door-and-Go-Through
    Travel to Start-Location
    END
```

So far so good. We activate our program and Uranus obediently begins to trundle down the hall, counting doors. It correctly recognizes the first one. The second door, unfortunately, is decorated with some garish posters and the lighting in that part of the corridor is poor, so our experimental *Door-Recognizer* fails to detect it. The *Wall-Follower*, however, continues to operate properly and Uranus continues on down the hall, its door count short by one. It recognizes door three, the one we had asked it to go through, but thinks it is only the second and so continues. The next one is recognized correctly, and it's open. The program, thinking this is the third door, faces it and proceeds to go through. This fourth door, sadly, leads to the stairwell, and poor Uranus, unequipped to travel on stairs, is in mortal danger.

Fortunately, there's a process in our concurrent programming system called *Detect-Cliff*. It is always running, checking ground position data posted on the blackboard by the vision processes; it also requests sonar and infrared proximity checks on the ground. It combines the data, perhaps with a high, a priori expectation of finding a cliff when operating in dangerous areas, to produce a number that indicates the likelihood of a drop-off somewhere nearby.

A companion process, *Deal-with-Cliff*, also running continuously but with low priority, regularly checks this number and adjusts its own priority on the basis of it. When the cliff probability variable becomes high enough, the priority of *Deal-with-Cliff* will exceed the priority of the current process in control (*Go-Fetch-Cup* in our example), and *Deal-with-Cliff* takes over control of the robot. A properly written *Deal-with-Cliff* will then proceed to stop or greatly slow down the movement of Uranus, increase the frequency of sensor measurements of the cliff, and slowly back Uranus away from it as soon as it has been identified and located.

Now, there's a curious thing about this sequence of actions. A person seeing them, not knowing about the internal mechanisms of the robot, might offer this interpretation: "First the robot was determined to go through the door, but then it noticed the stairs and became so frightened and preoccupied it forgot all about what it had been doing." Knowing what we do about the programming of the robot, we might be tempted to scold this poor person for using such sloppy anthropomorphic concepts as determination, fear, preoccupation, and forgetfulness in describing the actions of a machine. We *could* do so, but it would be wrong.

I think the robot *would* come by the emotions and foibles described as honestly as any living animal. An octopus in pursuit of a meal can be diverted by hints of danger in just the way Uranus was. An octopus also happens to have a nervous system that evolved entirely independently of our own vertebrate version. Yet most of us feel no qualms about ascribing concepts like passion, pleasure, fear, and pain to the actions of the animal.

We have in the behavior of the vertebrate, the mollusk, and the robot a case of convergent evolution. The needs of the mobile way of life have conspired in all three instances to create an entity that has modes of operation for different circumstances, and that changes quickly from mode to mode on the basis of uncertain and noisy data prone to misinterpretation. As the complexity of the mobile robots increases, I expect their similarity to animals and humans will become even greater.

Among the natural traits I see in the immediate roving-robot

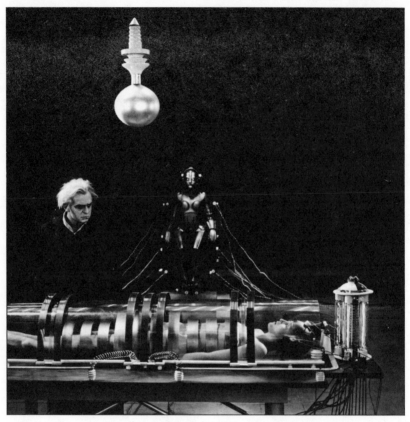

To imbue a machine with the capability to roam at will is,
in the estimate of some, to give it protohuman characteristics.

horizon is parameter-adjustment learning. A precision mechanical arm in a rigid environment can usually have its self-image adjusted once; it's a permanent adjustment. A mobile robot bouncing around in the muddy and rocky world will continually suffer from dirt buildup, tire wear, frame bends, and mounting bracket slips that mess up accurate, a priori models. For instance, our present software for the visual obstacle course has a camera calibration phase in which the robot parks itself precisely in front of an exact grid of spots so that a program can determine a function that corrects for distortions in its camera optics. This allows other programs to make precise measurements of visual angles in spite of distortions in the camera lens. We have noticed that our present code is very sensitive to miscalibrations, and we're working on a method to continuously calibrate the cameras from the images perceived on normal trips through clutter. With such a procedure in place, a bump that slightly shifts one of the robot's cameras will no longer cause systematic errors in its navigation. Animals seem to tune most of their nervous systems with processes of this kind, and such accommodation may be a precursor to more general kinds of learning.

Perhaps more controversially, I see the beginnings of self-awareness in the robots. All of our control programs have internal representations, at varying levels of abstraction and precision, of the world around the robot and of the robot's position within that world. The motion planners work with these world models in considering alternative future actions for the robot. If our programs had verbal interfaces, we could ask questions that would elicit answers, such as: *I turned right because I didn't think I could fit through the opening on the left.* As it is, we get the same information in the form of pictures drawn by the programs.

There may seem to be a contradiction in the various estimations of the speed of computers. Once billed as "giant brains," computers can do some things, like arithmetic, millions of times faster than human beings. "Expert systems" doing qualitative reasoning in narrow problem-solving areas run on these computers approximately at human speed. Yet it took such a computer five hours to simply drive the Cart across a room, down to an hour for Neptune. How can such differences be reconciled?

The human evolutionary record provides the clue. While our

sensory and muscle-control systems have been in development for *a billion* years and common sense reasoning has been honed for probably about *a million*, really high-level, deep thinking is little more than a parlor trick, culturally developed just *a few thousand* years ago. A few humans, operating largely against their natures, can learn this trick. As with Samuel Johnson's dancing dog, what is amazing is not how well it is done, but that it is done at all.

Computers can challenge humans in *intellectual* areas where humans perform inefficiently, because they can be programmed to carry on much less wastefully. An extreme example is arithmetic, a function learned by humans with great difficulty, but which is instinctive to computers. These days an average computer can add a million large numbers in a second, which is more than a million times faster than a person, and with no errors. (And yet, one hundred-millionth of the neurons in a human brain, if reorganized into an adder using switching-logic-design principles, could sum a thousand numbers per second. If the whole brain were organized this way, it could do sums one hundred thousand times faster than the computer!)

Computers do not challenge humans in *perceptual and control* areas because these billion-year-old functions are carried out by large portions of the nervous system operating as efficiently as the hypothetical neuron adder above. Present-day computers, however efficiently programmed, are simply too puny to keep up. Evidence comes from the most extensive piece of reverse engineering yet done on the vertebrate brain, the functional decoding of some of the visual system by David H. Hubel, Torsten N. Weisel, and their colleagues at MIT.

The vertebrate retina's 20 million neurons take signals from a million light sensors and combine them in a series of simple operations to detect things like edges, curvature, and motion. The image thus processed is sent to the much bigger visual cortex in the brain. Assuming the visual cortex does as much computing for its size as the retina, we can estimate the total capability of the system. The optic nerve has 1 million signal-carrying fibers, and the optical cortex is a thousand times thicker than the layers of neurons that do the basic retinal operation. The eye can process 10 images/second, so the cortex handles the equivalent of 10,000 simple retinal operations a second, or 3 million/hour.

An efficient program running on a typical computer can do the equivalent work of a retinal operation in about two minutes, for a rate of 30/hour. Thus, seeing programs on present-day computers seem to be 100,000 times slower than vertebrate vision. The whole brain is about 10 times larger than the visual system, so it should be possible to write real-time human equivalent programs for a machine 1 million times more powerful than today's medium-size computer. Even today's largest supercomputers are about 1,000 times slower than this desiratum. How long before our research medium is rich enough for full intelligence?

Since the 1950s, computers have gained a factor of 1,000 in speed-per-constant-dollar every decade. There are enough developments in the technological pipeline to continue this pace for the foreseeable future.

The processing power available to AI programs has not increased proportionately. Budget increases have been spent on convenience features—operating systems, time-sharing, high-level languages, compilers, graphics, editors, mail systems, networking, personal machines, etc.—and have been spread more thinly over ever greater numbers of users. I believe this hiatus in the growth of processing power explains the disappointing pace of the development of AI in the past fifteen years, but nevertheless it represents a good investment. Basic computing facilities are now widely available, and—thanks largely to the initiative of the instigators of the Japanese Supercomputer and Fifth Generation Computer projects—attention worldwide is focusing on the problem of processing power for AI.

The new interest in computing power should ensure that AI programs share in the thousandfold-per-decade increase from now on. This puts the time for human equivalence at twenty years. Since the smallest vertebrates, like shrews and hummingbirds, produce interesting behavior with nervous systems one ten-thousandth the size of a human's, we can expect fair motor and perceptual competence in less than a decade. By my calculations and impressions, present robot programs are now similar in power to the control systems of insects.

Some principals in the Fifth Generation Project have been quoted as planning "man-capable" systems in ten years. I believe this more optimistic projection is unlikely, but not impossible. The fastest present and nascent computers, notably the supercomputers Cray X-MP and Cray 2, compute at 10^9 operations/second, only they do it 1,000 times too slowly.

As computers become more powerful and as research becomes more widespread, the rate of visible progress should accelerate. I think artificial intelligence via the "bottom-up" approach of technologically recapitulating the evolution of mobile animals is the surest bet, because the existence of independently evolved intelligent nervous systems indicates that there is an incremental route to intelligence. It is also possible, of course, that the more traditional, specialized "top-down" approach, which starts with the narrow problem-solvers of today and goes into the much broader and harder areas of learning, common sense reasoning, and perceptual acquisition of knowledge, will achieve its goals as computers become large and powerful enough and the techniques are mastered. Most likely both approaches will make enough progress that they can effectively meet somewhere in the middle, for a grand synthesis into a true artificial sentience.

This artificial person will have some interesting properties. Its high-level reasoning abilities should be astonishingly better than a human's—even today's puny systems are much better in some areas—but its primitive, low-level perceptual and motor abilities will be comparable to ours. Most interestingly, it will be highly malleable, both on an individual basis and from one of its generations to the next. And it will quickly become *cheap*.

What will happen when these cheap machines can replace humans in any situation? What will *I* do when a computer can write this chapter and do research better than I can? These questions figure in some occupations now, but they will affect everybody in a few decades.

By design, machines are our obedient and able slaves. But intelligent machines, however benevolent, could threaten our existence because they will be alternative inhabitants of our ecological niche. Even machines merely *as* clever as human beings will have enormous advantages in certain competitive situations. They cost less to build and maintain, so more of them can

The lonely job of night watchman may one day be taken over
by sentry machines such as this one by Advanced Robotics, Inc.

be put to work with given resources. They can be optimized to do their jobs, and programmed to work tirelessly.

Intelligent robots will have even greater advantages away from our usual haunts. Very little of the known universe is suitable for unaided humans. Only by being surrounded with massive machinery can we survive in outer space, on the surfaces of the planets, or on the sea floor. Smaller, intelligent but unmanned devices will be able to do what needs to be done more cheaply. The Apollo project put people on the moon for *$40* billion. Viking landed machines on Mars for *$1* billion. If the Viking landers had been as capable as humans, their multiyear stay would have told us much more about Mars than the Apollo's crew found out about the moon.

As if this weren't bad enough, the very pace of technology presents an even more serious challenge. We evolved with a leisurely hundred million years between significant changes. Machines are making similar strides, but doing it in only decades. The rate will quicken further as armies of cheap machines are put to work as programmers and engineers, with the task of improving the software and hardware that makes them what they are. The successive generations of machines produced in this way will be increasingly smarter and cheaper. There is no reason to believe that human equivalence represents any sort of upper bound. When pocket calculators can out-add humans, what will a really big computer be like? We will simply be outclassed.

Then why rush headlong into the intelligent machine era? Wouldn't any sane human try to delay things as long as possible? The answer is obvious, if unpalatable at first glance. Societies and economies are as surely subject to evolutionary pressures as biological organisms. Failing social systems do wither and die, eventually to be replaced by more successful competitors. Those that can sustain the most rapid expansion dominate sooner or later.

We all compete with each other for the resources of the accessible universe. If automation is more efficient than hand labor, the organizations and societies that embrace it will be wealthier and better able to survive in difficult times, and to prosper in favorable ones. If the United States were to unilaterally halt technological development—an occasionally fashionable idea—

it would soon succumb either to the military might of the Soviets or to the economic success of its trading partners. Either way, the social ideals that led to the decision would become unimportant on a world scale.

If, by some evil and unlikely miracle, the whole human race decided to eschew progress, the long-term result would be almost certain extinction. After all, the universe is one random event after another. Sooner or later an unstoppable virus deadly to humans will evolve, or a major asteroid will collide with the Earth, or the sun will go nova, or we will be invaded from the stars, or a black hole will swallow the galaxy.

The bigger, more diverse, and competent a culture is, the better it can detect and deal with external dangers like these. By growing sufficiently rapidly it has a finite chance of surviving forever. The human race will expand into the rest of the solar system soon, and space colonies will be part of it. But the economics of automation in space will become very persuasive even before machines achieve human competence.

I visualize immensely lucrative self-reproducing robot factories installed on the asteroids. Solar-powered machines would prospect and deliver raw materials to huge, automatic processing plants. Metals, semiconductors, and plastics produced there would be converted by robots into components that would be assembled into other robots and parts for more plants. Machines would be disassembled and recycled as they broke down. If the reproduction rate is higher than the wear-out rate, the system will grow exponentially. A small fraction of the output of materials, components, and whole robots could make someone very, very rich.

The first space industries will be more conventional. Raw materials purchased from Earth or from human space settlements will be processed by human-supervised machines and sold at a profit. The high cost of maintaining humans in space ensures that there will always be more machinery per person there than on Earth. As machines become more capable, the economics will favor an ever higher machine/people ratio. Humans will not necessarily become fewer, but the machines will multiply faster.

When humans become unnecessary in space industry, the physical growth rate of the machines will climb. When machines reach and surpass humans in intelligence, the intellectual growth rate will rise similarly. The scientific and technical discoveries of superintelligent mechanisms will be applied to making themselves smarter still. The machines, looking quite unlike the machines we know, will explode into the universe, leaving us behind in a figurative cloud of dust. Barring prior claims, our intellectual, but not genetic, progeny will inherit the universe.

This may not be as bad as it sounds, since the machine civilization will certainly take along everything we consider important, including the information in our minds and genes. Real, live human beings, a whole human community, could be reconstituted, if needed. Since we are biologically committed to personal death, immortal only through our children and our culture, shouldn't we rejoice to see that culture become as robust as possible?

Yet some of us have very egocentric world views. We don't take kindly to being upstaged by our creations. We anticipate the discovery, within our lifetimes, of ways to extend human lifespans, and we can look forward to a few eons of exploring the universe. How can we, personally, become full, unhandicapped players in this new game?

Genetic engineering is one option. Successive generations of human beings could be genetically redesigned by mathematics, computer simulations, and experimentation, like airplanes and computers are now. But all this would be building robots out of protein. And away from Earth, protein is not an ideal material. It's stable only in a narrow temperature and pressure range, is sensitive to high-energy disturbances, and is not receptive to many construction techniques and components.

What's really needed is a process that gives an individual all the advantages of the machines, at small personal cost. Transplantation of human brains into manufactured bodies has some merit, because the body can be matched to the environment. It does nothing about the limited and fixed intelligence of the brain, which the artificial intellects will surpass. And so . . .

You are in an operating room. A robot brain surgeon is in attendance. Near you is a potentially human-equivalent computer, dormant for lack of a program to run. Your skull, but not your brain, is anesthetized. You are fully conscious. The surgeon opens your brain case and peers inside. Its attention is directed at a small clump of about a hundred neurons somewhere near the surface. It determines the three-dimensional structure and chemical makeup of that clump nondestructively with high-resolution 3-D NMR holography, phased-array radio encephalography, and ultrasonic radar. In seconds, it writes a program that mimics the behavior of the clump, and then starts the program running on a small portion of the computer next to you.

Fine connections are run from the edges of the neuron assembly to the computer, providing the simulation with the same inputs as the neurons. You and the surgeon check the accuracy of the program. After you are satisfied, tiny relays are inserted between the edges of the clump and the rest of the brain. (Initially these leave the brain unchanged, but on command they can engage the simulation in place of the clump.) A button that activates the relays when pressed is placed in your hand. You press it, release it, and press it again. There should be no difference. As soon as you are satisfied, the simulation connection is established firmly, and the now unconnected clump of neurons is removed. The process is repeated over and over for adjoining clumps, until bit by bit the entire brain has been replaced by a machine. Occasionally several clump simulations are combined into a single equivalent but more efficient program. Though you have not lost consciousness, or even your train of thought, your mind (some would say soul) has been removed from the brain and transferred to a machine.

In a final step, your old body is disconnected. The computer is installed in a shiny new one, in the style, color, and material of your choice. You are no longer a cyborg halfbreed; your metamorphosis is complete.

For the squeamish, there are others ways to do the transfer. The high resolution brain scan could be done in one fell swoop, without surgery, and a new you made ("While-U-Wait"). Some will object that this instant process makes only a copy, that the real you is still in your old body to dispose of properly. This is an understandable misconception based on the intimate association of a person's identity with a particular, unique, irreplaceable piece of flesh. Once the possibility of mind transfer is accepted, however, a different notion of life and identity becomes possible. You are not dead until the last copy is erased, and a faithful copy is exactly as good as the original.

If even this technique is too invasive for you, imagine a more psychological approach. A kind of pocket computer (perhaps shaped and worn like glasses) is programmed with the universals of human mentality, your own genetic makeup, and details of your life easily accessible. It carries a program that makes it an excellent mimic. You carry this computer with you through the prime of your life. It diligently listens and watches, and perhaps monitors your brainwaves, and learns to anticipate your every move and response. Soon it is able to fool your friends on the phone with its convincing imitation of you. When you die, it is installed in a mechanical body to smoothly and seamlessly take over your life and responsibilities.

Still not satisfied? Since you are a vertebrate, there is another option that combines some of the features of all the methods I've described. The vertebrate brain is split into two hemispheres connected by a very large bundle of nerve fibers called the corpus callosum. When brain surgery was new, doctors discovered that severing this connection cured some forms of epilepsy. An astounding aspect of the operation was that the patient had no apparent side effects.

Now, the corpus callosum is a bundle far thicker than the optic nerve or even the spinal cord. Cut the optic nerve and the victim is utterly blind; sever the spinal cord and the body goes limp. Slice the huge cable between the hemispheres and nothing happens? Well, not quite. It was noted that patients who had this surgery were unable, when presented with the written word *brush*, for instance, to pick out the object in a collection of others, using the left hand. The left hand would wander uncertainly from object to object, seemingly unable to decide which object was *brush*. When asked to do the same task with the right hand, the choice was quick and unhesitating. Sometimes, in the left-handed version of the task, the right hand, apparently in exasperation, reached over to guide the left to the proper location. Researchers observed other such quirks involving spatial reasoning and motor coordination.

The explanation is that the callosum is the main communications channel between the brain hemispheres. It has fibers running to every part of the cortex on each side. The brain halves, however, are fully able to function separately and call on this channel only when a task involving coordination becomes nec-

essary. We can postulate that each hemisphere has its own prior-
ities; that the other can request, but not demand, information
or action from it; and that each must be able to operate effectively if the other chooses not to respond, even when the callosum is intact. The left hemisphere handles language and controls
the right side of the body. The right hemisphere controls the left
half of the body, and without the callosum the correct interpretation of the letters *b r u s h* could not be conveyed to the controller of the left hand.

But what an opportunity. Suppose we sever your callosum
but also connect a cable from an external computer to both
severed ends. If we understand the human brain well enough,
this external computer can be programmed not only to pass
information, but also to monitor the traffic between the two
halves. Like the personal mimic, it can teach itself to think like
them. After a while it could insert its own messages into the
stream, becoming an integral part of your thought processes. In
time, as your original brain fades away from natural causes, it
can smoothly take over the lost functions. Ultimately your mind
finds itself in the computer. With advances in high-resolution
scanning, it may even be possible to have this effect without
messy surgery—you would just wear some kind of helmet or
headband.

Whatever method you choose for brain exchange, when the
process is complete the advantages become apparent. Your computer has a control labeled SPEED. It had been set to SLOW, to
keep it synchronized with the old brain, but now that the brain
is gone you change it to FAST. You can communicate, react, and
think a thousand times faster. But wait, there's more!

The program in your machine can be read out and altered,
letting you conveniently examine, modify, improve, and extend
yourself. The entire program may be copied into similar machines, giving two or more thinking, feeling versions of you. You
may choose to move your mind from one computer to another
more technically advanced, or more suited to a new environment. The program can also be copied to some future equivalent
of magnetic tape. If the machine you inhabit is fatally clobbered,
the tape can be read into a blank computer, resulting in another
you, minus the experiences since the copy. With enough copies,
permanent death would be very unlikely.

In the form of a computer program, your mind can travel over information channels. A laser could beam it from one computer to another across great distances. If you found life on a neutron star and wished to make a field trip, you might devise a way to build a neutron computer and robot body on its surface, then transmit your mind to it. Nuclear reactions are a million times quicker than chemistry, so with the neutron computer you can probably think that much faster. It can act, acquire new experiences and memories, then beam its mind back home. If you wished, your original body could be kept dormant during the trip to be reactivated with the new memories when the return message arrived.

Alternatively, the original might remain active. There would then be two separate versions of you, with different memories for the trip interval. The two sets of memories can be merged, if mind programs are adequately understood. To prevent confusion, memories of events would indicate in which body they happened. Merging should be possible not only between two versions of the same individual but also between different persons. Selective mergings, involving some of the other person's memories, and not others, would be a very superior form of communication, in which recollections, skills, attitudes, and personalities could be rapidly and effectively shared.

Your new body will be able to carry more memories than your original biological one, but the accelerated information explosion will make it impossible to lug around all of civilization's knowledge. You will have to pick and choose what your mind contains at any one time. There will often be knowledge and skills available from others superior to your own, and the incentive to substitute those talents for yours will be overwhelming. In the long run you will remember mostly other people's experiences, while memories you originated will be floating around the population at large. The very concept of *you* will become fuzzy, replaced by larger, communal egos.

Mind transferral need not be limited to human beings. Earth has other species with brains as large, from dolphins (our encephalic equals) to elephants, whales, and giant squid, whose brains are up to twenty times as big as ours. Translation between their mental representation and ours is a mere technical problem comparable to converting our minds into a computer program. Our culture could be fused with theirs; we could incorporate each other's memories and the species' boundaries would fade. Nonintelligent creatures could also be included in the data banks. Even the simplest organisms might contribute a little information from their DNA. In this way our future selves will benefit from all the lessons learned by terrestrial biological and cultural evolution. This is a far more secure form of storage than the present one, where genes and ideas are lost when the conditions that gave rise to them change.

Our speculation ends in a supercivilization, the synthesis of all solar system life, constantly improving and extending itself, spreading outwards from the sun, converting nonlife into mind. There may be other such bubbles expanding from elsewhere. What happens when we meet? Fusion of us with them is a possibility, requiring only a translation scheme between the memory representations. This process, possibly already occurring elsewhere, might convert the entire universe into an extended thinking entity, a probable prelude to greater things.

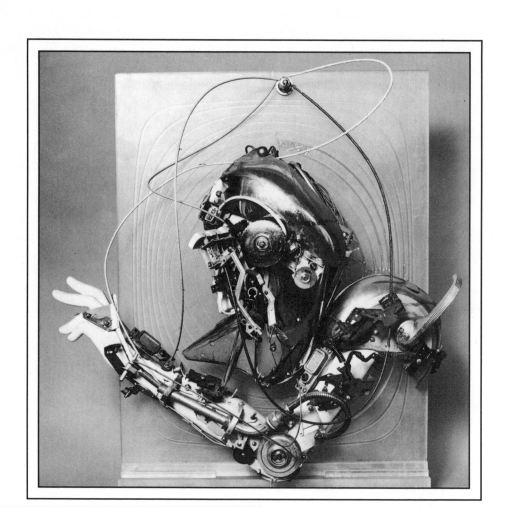

The Birth of the Cyborg

Robert A. Freitas

The emerging symbiosis between man and machine will pervade our lives more deeply than many of us have dreamed. Machines are now being physically married to the human body. In time the distinction between man and machine will begin to blur, giving rise to hybrid cybernetic organisms, or cyborgs, the ultimate man-machine symbiosis.

In a famous science-fiction tale by Robert A. Heinlein, a wealthy but crippled young man named Waldo invents amazingly deft remote-controlled arms and hands that imitate and amplify the power of his own weakened limbs. Waldo creates dozens of mechanical hands—some human size, others microscopic, even one with a 20-foot span. He passes his time operating factories on Earth, effortlessly from space, in zero gravity.

Waldo was a fictional character of the 1940s, but today fiction is fast becoming fact. Remote manipulators were first used in the 1940s to handle dangerous nuclear materials from a safe distance. Since then, these robotlike devices have been used to explore space and the sea, to troubleshoot stricken nuclear power plants, to rehabilitate the disabled, to assist in police work, and to perform surgery. There are also military applications. Remotely piloted spy planes fly missions regularly around the world; man-amplifiers, walking tanks, and thought-controlled weaponry have been and are being studied for future use.

Remote-controlled systems, also known as teleoperators, are incredibly powerful tools because they can manipulate the environment in a manner vastly beyond the range of normal human capabilities. Much as the computer extends the power of our minds, teleoperators project man's physical reach in many new and unexpected ways. They can amplify human motions to a gigantic scale or subdue them to delicately precise micromovements; apply inhumanly powerful or subtly controlled forces; initiate physical movement more swiftly or more slowly than people can; and multiply human effort by "slaving" many machines choreographed by a single human "master."

Today's planners are looking beyond simple teleoperation to a new concept called telepresence. In telepresence, the mechanical system at the worksite closely corresponds to human senses and limbs. Back at the control station, the human operator receives high-quality, complete sensory feedback that gives him the feeling of actually being at the faraway worksite. The sensory and kinesthetic illusion of remote presence is so complete that the human controller reacts quickly and correctly to the signals from the distant environment, taking advantage of his or her learned reactions and human decision-making abilities.

Someday, telepresence may make possible a remotely manned economy. In such an economy, most physical labor would be performed by teleworkers—people operating machinery hundreds or even thousands of miles away, using home control stations in local community work clubs. In their leisure time, these people may vacation as teletourists or keep in shape via telesports.

The emerging symbiosis between man and machine will pervade our lives more deeply than many of us ever dreamed. Machines are now being physically married to the human body. By the turn of the century, bionic limbs, senses, and organs—each as good or better than the originals—will extend the human lifespan, increase our physical endurance and stamina, and make the human body stronger and more capable. In time the distinction between man and machine will begin to blur, giving rise to hybrid cybernetic organisms, or cyborgs, the ultimate man-machine symbiosis.

It was the U.S. military who first became interested in teleoperators in the late 1950s and early 1960s. The proposed "man-amplifier," a powered exoskeleton, could conceivably transform

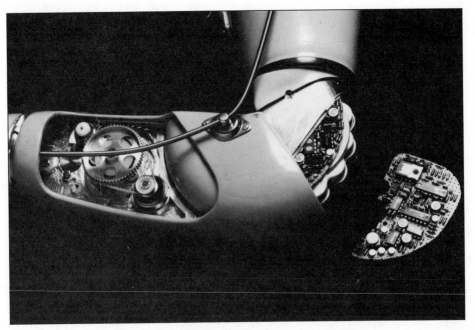

This intelligent prosthetic arm, from the University of Utah, converts fine muscle contractions into delicate limb movement.

an ordinary combatant into a supersoldier. After exploring a few possibilities, army, navy, and air force engineers joined in a development program with General Electric Company called HARDI-MAN (Human Augmentation Research and Development Investigation). Their goal was to create "superman suits."

The Hardiman concept is a mechanical exoskeleton worn by a human operator. To use it, the man slips into a harness called the master (the control system), encased in an outer mechanical structure called the slave. To get the slave to move an arm upward, the operator simply raises his arm. This action is detected by sensors in the master harness and relayed to the slave. To make the machine stoop down and pick up a load, the operator simply performs the motion in the natural way. The device would amplify a man's strength 25 times. Wearing this suit, you could lift a 1,500-pound load (which would feel like only 60 pounds) to a height of 6 feet in 5 seconds, then carry it 25 feet in 10 seconds. Any number of terminal devices could be attached. Hands could become high-speed electric drills or grippers; different sets of feet could be attached for travel over smooth or rugged terrain.

One Hardiman would consume as much power as a small automobile needing up to 60 horsepower during peak loads and 15 horsepower standing still. Mechanical supermen would wear lightweight gasoline engines or gas turbines on their backs, plus enough fuel for several hours. By 1969, General Electric's project leader Ralph S. Mosher confidently predicted: "The Hardiman concept will be used for bomb-loading, underwater construction, and many materials-handling tasks. The operator is able to react in such a natural manner that he subconsciously considers the machine as part of himself." The concept was proven in the laboratory, and working arms and legs were built, but to the best of my knowledge, a complete superman suit was never actually constructed.

While Hardiman was primarily a materials-handling and repair device, the same idea could lead to a kind of intelligent battle armor for individual soldiers. A controllable, powered framework surrounding a soldier might amplify his strength and serve as a protective shell. The soldier in effect becomes a walking tank, able to carry a variety of heavy armament without losing the versatility and mobility of an individual. The best fictional description of intelligent battle armor was penned by Robert Heinlein in his 1959 novel *Starship Troopers*. His mobile infantry of the future wears 1-ton powered suits. Each suit has strong but lightweight armor, powerful weaponry, sensors, and small jets for house-high jumps, all with feedback to reflect the soldier's every movement. Heinlein's protagonist applauds his powered armor: "You just wear it and it takes orders directly from your muscles and does for you what your muscles are trying to do. This leaves you with your whole mind free to handle your weapons and notice what is going on around you—which is *supremely* important to an infantryman who wants to die in bed."

The idea of building manned walking machines is not exactly new, either. A. C. Hutchinson, a British engineer, proposed a 1,000-ton walking tank in 1940. And in 1954, a Professor Bernhard of Rutgers University proposed a grasshopperlike jumping vehicle to the military. Patents for various types of walking devices date back more than a hundred years. "Compared with the versatility of man," says GE's expert Ralph Mosher, "the things a machine can be programmed to do are extremely limited. For example, it would be impractical to construct a machine able to make its way by itself through sand, mud, rocks, swamp, and forest—and to do this without knowing what was coming up next. But a man can do this, and if you join man and machine, using the best capabilities of each—man's brains and the machine's great strength—then a machine can do it, too."

There had always been some fear that an operator standing or walking on 12-foot legs and attempting to do tasks with 6-foot arms might become confused and disoriented, or be thrown off balance. To find out, General Electric engineers built the Pedipulator Balance Demonstrator, an 18-foot-high, two-legged balance-testing machine, at the U.S. Army Tank and Automotive Center. Operators were strapped into a harness in a cage atop the teetering frame, their heads 15 feet above the floor. They easily kept from falling over by balancing themselves with hip and ankle control movements mimicked by the device. After obtaining these results, Mosher reflected on the future: "It is conceivable that a 20- or even a 50-foot-tall Hardiman can be built with the operator inside as the 'brain' or controller. A pedipulator with legs 18 feet long could move over the ground at the rate of about 35 miles per hour if its human operator moved his legs at a six-miles-per-hour gait." Of course, even such impressive walkers are not invincible. As the drivers of the Imperial Walkers discovered to their dismay in George Lucas's film *The Return of the Jedi*, two-legged war machines can fall or be tripped.

Pursuing a more practical system, the army commissioned a demonstration Quadrupedal Walking Machine to carry 500 pounds over rough terrain. In 1969, the result of the million-dollar project was fueled up with gas, and Mosher danced it through its paces. At first glance the walking truck looked like a four-legged robot, a gleaming metal monster standing 11 feet high and weighing 3,000 pounds. It moved about on its hydraulic haunches with surprising fluidity and grace, suggesting a well-trained circus elephant. Mosher, tucked within the beast's entrails, was its brain and nervous system.

Strapped inside a control harness, electronically and hydraulically linked to the machine's appendages, and equipped with sophisticated force reflectors and actuators, Mosher had only to make simple crawling motions to move the quadruped along. With the flick of a wrist he kicked 175-pound railroad ties out of his path as if they were toothpicks. Slamming his foot down shook the building with a 1,500-pound wallop from the machine's corresponding hind leg. The man-machine relationship, Mosher later recalled, "is so close that the experienced operator begins to feel as if those mechanical legs are his own. You imagine you are actually crawling along the ground on all fours— but with incredible new strength."

Walking machine research continues today worldwide. Projects at the USSR's Moscow State University and, in Japan, a $140-million cooperative effort between government and business represent the most extensive overseas development efforts. In the United States, probably the most advanced unmanned walking machine is the radio-controlled ODEX I "functionoid" manufactured by Odetics of Anaheim, California. The 370-pound spiderlike robot stands on six legs and can squat low to pick up heavy loads or stand up tall and thin to tiptoe through doorways and narrow halls. Legs can be moved individually or commanded to follow programmed gaits. ODEX I can stroll along at 4 miles per hour and can hoist a ton. In one recent demonstration it stepped off the back of a pickup truck, then turned around, lifted the truck's rear end off the ground, and walked the vehicle through a 90-degree turn across the floor. Future versions will incorporate manipulator arms, greater durability, and more on-board intelligence.

As for manned walkers, Ivan Sutherland at Carnegie-Mellon University in Pittsburgh is working on a six-legged, hydraulically driven crawling machine. An 18-horsepower gasoline engine provides power, and hydraulic actuators move the legs under the control of a built-in microprocessor guided by leg sensors. The computer simultaneously regulates the machine's gait, keeps it from falling over, distributes correct loads to each leg, and makes sure no leg is driven past its design limits. To ride in Sutherland's machine, a person sits in back where he can see the legs. Mincing along at a cautious 2 miles per hour, the driver steers by causing the legs on one side to move faster than those on the other. He can also adjust the attitude of the body and its height above the ground, and make it roll left or right or pitch forward or backward while standing. Since at least three of its six legs always touch the ground, the machine doesn't need a sense of balance.

An even more ambitious walking machine is the Adaptive Suspension Vehicle (ASV) now being built at Ohio State University in Columbus. The six-legged, two-and-a-half-ton metal behemoth measures 15 feet long and 5 feet wide, and stands 10 feet high while walking. It can carry a single operator in an enclosed cabin, plus a 500-pound cargo. Its cruising speed is 5 miles per hour, but it can sprint at 8 miles per hour. A 900-cc motorcycle engine drives an energy storage flywheel that allows the ASV to scramble up a 60-degree incline or step across a 6-foot ditch. The human operator commands direction and speed using a joystick. Unlike the Carnegie-Mellon crawler, the Ohio system uses a computerized terrain-scanning system that surveys the ground ahead of the machine and chooses the best footholds. An infrared optical radar beam sweeps the surface 30 feet ahead, sensing obstacles as small as one half inch wide. This information—together with data from the proximity and force sensors on its feet, its gyros, and its accelerometers—is fed to the coordination computer, which selects the correct path over normal terrain.

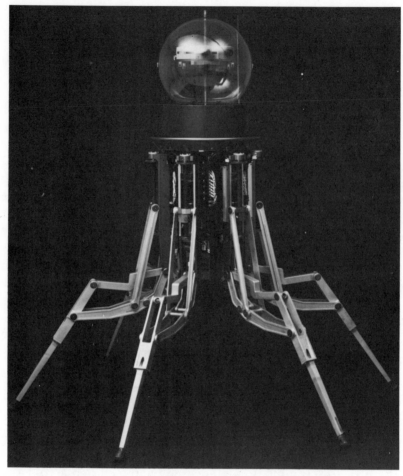

This six-legged "functionoid" could be the telepresence
drone of the future. With a TV camera in its glass-domed head
and six-legged locomotion, it could be directed to go over
almost any terrain and squeeze through the narrowest of openings.

All these machines lack the ability to balance dynamically, which Sutherland believes is crucial to future systems. "Crawlers will ultimately be replaced by machines with fewer legs that can balance," says Sutherland. "Mastery of balance is the key to building high-mobility machines that can walk and run."

In the late 1940s, radiochemists were faced with a novel problem: how to untangle the hundreds of radioactive fission products found in spent nuclear fuel, but from behind several feet of concrete and lead. Long tongs weren't dexterous enough and glove boxes were unsafe. In 1948, Ray Goertz and others at the Argonne National Laboratory built the first mechanical master-slave manipulator with force feedback that enabled operators to feel what they were touching. The descendants of these early systems are in wide use throughout the world today.

The future belongs to unmanned systems like Mobot, an early track-drive vehicle supporting two 100-pound-capacity manipulators and grippers with 200-pound handshakes. The manipulators, plus two independent camera booms, an overhead 1,000-pound jib crane and a forward 1,000-pound forklift were powered by 8-hour batteries. The largest system ever built is the air force's unmanned Mobile Remote Manipulator Unit (MRMU). The MRMU travels at 10 miles per hour, and climbs steep slopes; it has two 19-foot, 600-pound-capacity manipulator arms and, between them, four cameras on a boom. Converted from a full-tracked army cargo carrier, the MRMU is about the size and shape of a tank and evokes visions of future wars fought entirely by remote-controlled roboweapons.

Today, research in unmanned manipulation vehicles aims for high versatility. The MF3-E manipulator vehicle, currently in use by the West German Nuclear Emergency Brigade, is an excellent example. The half-ton robot is controlled by a pair of stereoscopic television cameras with zoom lenses; it has two searchlights, stereo microphones, a gamma-ray dosimeter, a temperature sensor, and two master-slave arms with a 5-foot reach and a 50-pound capacity. The most remarkable feature of the MF3-E is its two swiveling drive-track pairs, which enable it both to roll flat on the ground like a tractor and also to stand up, climb stairs, walk, or kneel like a biped.

Unfortunately, most present-day remote manipulators are still controlled by pushbuttons, knobs, joysticks, and simple TV mon-

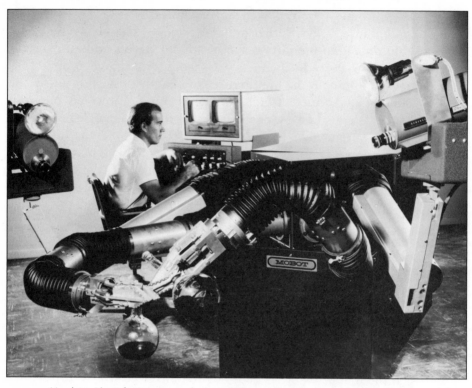

Hughes Aircraft Company designed this early remote-controlled worker,
a two-armed robot cart to handle "hot" nuclear materials.

itors. The movements required to manipulate the controls are
not natural ones, and the man-machine sensory links are poor.
Another big problem with mechanical master-slaves is that they're
uncomfortable to use. A pair I tried at Los Alamos Scientific
Laboratories many years ago tired my hands after a few minutes
of use. Researchers have been trying to improve this drawback
by developing computerized anthropomorphic (humanlike) controller and manipulator mechanisms.

An anthropomorphic teleoperator was developed in the early
1970s for the navy's Explosive Ordnance Disposal group by
MBAssociates. An old demo film shows a man wearing a gleaming, weblike sensor suit covering his upper torso and his right
arm and hand. A few feet away, a remarkably humanlike remote
hydraulic manipulator arm stacks blocks on a table and threads
a needle, closely matching the man's natural motions. The slave
arm reportedly could move as fast as the motions of any human
subject ever tested.

Proprioceptive (proprioception is the sense of limb position),
temperature, tactile, and pressure sensors have been developed
and are being improved, but the problem of how to reflect this
sensed data back to the operator is still not fully solved. Current
video display systems are cumbersome and artificial, as well.
To provide a true sense of reality and remote presence to the
operator, Dr. Kevin Corker and his colleagues in the UCLA Engineering Systems Department proposed a design for an ingenious "master glove" controller. Dr. Corker, now a member of
the technical staff of the Advanced Teleoperator Lab at NASA's
Jet Propulsion Laboratory, points out that people are adaptive
and can respond to anomalies. These abilities should be exploited, he says, not by distancing man from the machine, as in
supervisory control systems, but rather by "more tightly cou-
pling man and the manipulator."

Exoskeleton control suits should have extensive physical dis-
plays, Corker suggests. His master glove controller would provide proprioceptive feedback using countertorques at each
rotational joint. Loads distributed over the entire arm's surface
would be displayed as such, via inflatable fluid sacs arrayed over
arm and hand, each filled with hot or cold liquid to accurately
display slave arm temperature distributions. Forces would be
transmitted to each finger via torque motors, and a sense of

touch communicated to the operator by vibrotactile trans-
ducers. Electrocutaneous stimulation can provide sensations of
pressure, pain, and heat. "By immersing the operator so completely in this man-machine system," Corker says, "a remote
system may for the first time possess the adaptability of the
human being in unexplored environments and unstructured situations."

Such a sophisticated sensory display system has not yet been
built, partly because of the huge and complex computational
requirements. Still, JPL researchers are hard at work fabricating
a multiarticulated slave hand for use with a master glove controller. Corker is looking at end effectors with multiple modality
sensor systems for display to an operator. He is also investigating
how humans use various proprioceptive and other cues to accurately determine limb position and movement, and to sense
obstructions.

Most of our knowledge of our surroundings comes to us through
our eyes. A good visual feedback system is obviously essential
for true telepresence. The first real attempt at this was the Head-
sight remote surveillance system reported by Philco engineers
Charles Comeau and James Bryan in 1958. Comeau mounted a
television camera atop a building and attached its remote controls to a helmet, which he wore. Attached to the helmet was a
television screen, and every time Comeau moved his head the
camera followed his movement. The changing vista was projected onto the television screen and gave the helmet wearer the
feeling of standing on top of the building, looking around the
premises. The camera can swing full circle, but to reduce operator motion the engineers put a 2:1 ratio on the neck movement. Turning your head 30 degrees swivels the remote eye 60
degrees, giving a "rubberneck" effect. Unfortunately, the image
in that first test was weak and grainy, the field of view only 40
degrees wide, and resolution just one-tenth that of the human
eye.

A significantly improved system called foveal-HAT (Head-Aimed
Television) was pioneered in the late 1960s and early 1970s by
John Chatten of Telefactors Corporation in West Conshohocken,
Pennsylvania. The foveal-HAT image is a combination of two
television images derived from two separate cameras. One pic-
ture, the peripheral image, is about 70 degrees wide with one-

tenth human eye resolution. Superimposed over this is a second 8-degree foveal image, closely matching the field and acuity of straight-ahead, human-eye foveal vision. The images are mixed on a TV screen in a viewing hood worn by the operator. Chatten says the hood is so lightweight that you can swing your head a full half-circle in about a second.

To operate his system, Chatten hooked it up to a radio-controlled pickup truck. The operator's station received visual feedback from the remote driver through the foveal-HAT system; the operator also had a duplicate set of truck controls—steering wheel, brakes, accelerator, and gearshift—to which the actual truck controls were connected. William Bradley, an expert in teleoperator systems who helped develop the earlier Philco system, was asked to test-drive the remote vehicle.

Bradley stepped into the control cockpit and drove the truck "as if I'd been in it," he recalled. "I saw the skyline around me and saw that the truck was parked at one edge of a large field with a group of large trees, spaced about 50 feet apart, on the opposite side. I started the truck and made a left turn toward the trees. I found it easy to maneuver between and around the trees, then drove out to an open space where a set of flags had been set up to provide a sort of slalom course. It was easy to drive the course at 10 to 20 miles per hour." The vision system worked well, but Bradley was surprised by the lack of driving cues we usually take for granted. "When the vehicle made a sharp turn, I felt no acceleration. Also, it was a very smooth ride," he complained. "None of the bumps were conveyed to the driver."

Despite his successes, Chatten had difficulty finding financial support to pursue his systems beyond the development stage. The need for infusions of additional research dollars scared off potential users, and Chatten became discouraged. Still, progress continues at a slow pace.

The hazards of police, fire, and bomb-disposal work have spawned a generation of small, versatile, low-cost teleoperators that are already doing work too dangerous for people. One typical system is the $20,000 Canadian-built Pedsco Mark-3 unit, a bomb squad cop on wheels that can pick up parcels with its robot arm or handle a hose to put out automobile fires. A human cop, working controls on a remote dolly and seeing on a TV

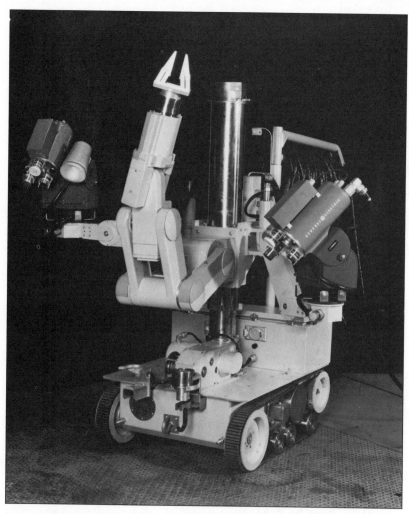

Long before the catastrophe of Three Mile Island, roboticists were trying to build atomic-materials-handling machines like this one.

screen what the Mark-3 sees through its 360-degree-pan video camera, can lower the arm on its pneumatic boom and remotely close the hand on any suspect package. New York City is using four Mark-3's to help answer some of its estimated 9,000 bomb scares per year, and more robots will soon be pressed into duty by the city's Emergency Service Division to respond to calls for everything from burning cars to hostage situations.

A lighter and more versatile system is AI Security's Ro-Veh (Remotely Operated Vehicle) which fits comfortably in the trunk of a small hatchback car. It travels on wheels or tracks at 1 mile per hour, and is powered through a reeled 300-foot control cable. The robot can open a car door, move through tall grass, crawl up stairs at a 45-degree incline, then position itself to millimeter accuracy—all by remote control. Ro-Veh's articulated boom arm comes with a wide assortment of attachments such as TV cameras, thermal image cameras, floodlights, x-ray equipment, a high-pressure water hose, loudspeaker, gripper, and armaments including a bank of slug-throwing bomb disrupters, a five-shot automatic shotgun, and an explosives dropper with a remote detonator.

The ocean is a dynamic environment that's beoming increasingly harsh for human divers as marine resource developers move from the shallow shelf waters into deeper waters in search of oil, gas, and mineral reserves. Remotely piloted, remotely controlled, unmanned, or robot vehicles can perform tasks such as inspecting the underwater rigging of oil drilling platforms at depths no human could withstand. There are now about twenty-five firms that operate undersea vehicles and about twenty manufacturers of these vehicles. Typical are the Hydro Products 2-foot-diameter "flying eyeball" RCV-225 and their newer 4-foot RCV-150 remote eye and light manipulator vehicle. For heavy-duty work, Hughes Aircraft Company's UNUMO underwater manipulator bristles with five snakelike manipulator arms, and is used in underwater oil well repair. Like most current systems, this underwater vehicle and its manipulators are controlled by a simple joystick, with the operator looking at a small TV monitor screen.

Studies at MIT show that as the manipulator becomes more manlike, the time needed to perform a task decreases and the complexity of the tasks that can be handled increases. According

Man-controlled machines, like this rifle-armed Ro-Veh,
can be made to rush in where wise men would never go.

to Thomas Sheridan of the MIT Sea Grant Program, if the tele-operator is designed to be anthropomorphic, the operator seems to identify his own body and his immediate environment with the remote vehicle and its environment. He identifies his own arm and his hand-control, or master arm, with the remote arm of the vehicle, and he identifies his head orientation with the orientation of the remote video camera. Sheridan calls this tendency of the operator to sense the teleoperator's orientation by relation to the orientation of his own body tele-proprioception. To better achieve this effect, Soviet experimenters have even placed operators in control cabs that pitch and roll as the remote vehicle pitches and rolls, thus providing direct inner ear balancing cues for tele-proprioception.

Monitoring the performance of robotic arms or manipulators is difficult when not enough information is relayed to the operator to tell him whether a task was done correctly. John Schneiter, an MIT doctoral candidate in the Department of Mechanical Engineering, is outfitting an underwater manipulator arm with tactile feedback that reports forces and patterns of touch. Touch-sensing is especially valuable in situations where the usual video link is impossible, such as in turbid or silt-filled water, or where the arm or environmental objects block the camera. With good touch sensing, a person could learn to recognize remote objects by feel and rescue items that have fallen out of sight.

Researchers have also employed undersea remote viewing systems. The French navy is pursuing this in its deep submergence teleoperator system, ERIC II, designed for "remote observation, investigation, and intervention" to 18,000-foot depths. The French have given special attention to "telesymbiotic" hardware. A TV camera system mounted on ERIC's front end is slaved to the surface operator's helmet and can follow all three head rotations and all three translational motions with a binocular foveal-HAT display. A pair of underwater microphones feed sound to stereo speakers in the helmet, and two dexterous force-feedback manipulator arms with touch-sensitive hands are under development.

Two distinct approaches to undersea automation are now being pursued. The first idea, called supervisory control, is that humans should give only "high-level commands," such as talking, to automated systems rather than operate manipulators directly. A good supervisory control system would understand a fixed vocabulary of commands or goals, and have a performance procedure for each one. The system could work interactively with its environment to accomplish these (possibly repetitive) goals without direct human intervention, according to proponents Thomas Sheridan of MIT and Robert Wernli of the Naval Ocean Systems Center (NOSC), San Diego.

The second idea, and the goal for the future, according to Wernli, is fully autonomous vehicles that can fulfill their missions without further intervention from any human. The ultimate powers of such machines, Wernli muses, will fall just short of the capabilities of HAL, the rebel system of the movie *2001: A Space Odyssey*. The prospect of becoming a junior partner in such a relationship worries many.

The second approach to marine automation is full telepresence, a path that may lead to a total symbiosis of man and machine. In marine telepresence, a mechanical system closely corresponds to human senses and limbs, building an illusion of presence at the distant underwater worksite. The human in control reacts quickly and naturally in the remote sea environment, taking advantage of a lifetime of learned reactions and decision-making abilities.

One of the best American marine telepresence systems is a package now being developed by David G. Stuart of NOSC-Hawaii, in Kailua. One major current application is remote, deep-sea, telepresence diving operations connected with the maintenance and repair of oil platforms. Stuart's system is controlled by an exoskeletal pair of arms worn by an operator. The system also has a mechanical spine and neck system, and can nearly duplicate human upper body movements. A black-and-white, head-aimed, helmet-mounted stereo vision system is slaved to all three rotations of the operator's head. Amazingly, while the NOSC system is one of the most advanced in existence, the technology consists entirely of off-the-shelf or decades-old hardware designed for other purposes.

In forty years of research and development, the basic technology of teleoperation and man-machine telesymbiosis has passed from dream to reality. It's not too early to begin thinking about the possibility of telework and what this would mean in each of our lives.

Arthur J. Critchlow of Mobility Systems in Santa Clara, California, has even proposed a nationwide teleoperator-based production system that would cost almost $700 million a year to operate. "We can use a computer to control a number of teleoperators remotely and have a human supervisor—a telesupervisor—override," he says. In his telework network, explains Critchlow, "there are 100 computers in a hundred locations around the country. There are exoskeletons on the people—sensors, displays, and teleoperator controls—20,000 sets of those. We figure that 20,000 telesupervisors, each controlling five teleoperators on the average, can run 100,000 teleoperators at remote terminals." The remotes would do useful work "either in domestic situations, or in unpleasant environments where people aren't willing to work."

Social mobility and labor patterns may be drastically altered by the advent of telework. Fewer people will need to physically travel "to work," but many may not want to stay home, either, and may invent "work clubs." Telelaborers can enter wider universes, working with others from diverse backgrounds. Persons with unusual talents can reach larger opportunities. Time-sharing multiple jobs will become more practical, and telejobs may be far more interesting and creative.

If sophisticated telepresence becomes widely available, could you get a telerobot to play sports for you? This could be an especially interesting prospect for highly dangerous activities you might not otherwise have the nerve to try—teleoperated boxing, racecar driving, parachuting, or mountain climbing. Telesports would let people feel reckless without risking personal harm.

Even relatively safe activities such as tennis and baseball could be teleoperated, resulting in faster and more challenging games. Already there are "track-and-field" events for man-operated machines—the Grand Prix, powerboat racing, tractor-pull contests, and radio-controlled model airplane and boat races. Using telepresence and variable-ratio force reflection, you could still work up a sweat and get exactly as much exercise as you wanted. You could teleoperate huge elephantlike or small dog-size robots, making possible whole new classes of sports experiences.

Teleoperation experts Edwin Johnsen and William Corliss speculate that some P. T. Barnum of the future may "fill parades and circus rings with giants, monsters, and robot gladiators that duel to the death. Indeed, combat by teleoperator might become a fad like 'crash' contests between jalopies. And what a status symbol a walking-machine golf caddy could be! To future generations no safari or mountain-climbing expedition may seem complete without teleoperators to clear the trail and carry supplies."

Even stranger possibilities present themselves. Telerobots could revive the dueling tradition—insulted ladies and gentlemen could square off their dueling robots at twenty paces. Thrillseekers might experience vicarious teledeath, committing simulated remote-suicide in robot war games or in destruction derbies, or by telecrashing an airplane into a mountain or telejumping off a cliff into the Grand Canyon. Convicted felons could be sent to teleprison, where they would be condemned to live telelives in a simulated hell or to reenact their own crimes a hundred times, but with themselves as the victim.

Perhaps a more pleasant prospect is teletourism—the use of robot remotes to do your leisure traveling for you. David Yates, a computer scientist at the National Physical Laboratory in London, calls these robots proxies. Yates suggests proxies should be roughly humanoid in shape and size, and be equipped with TV cameras, microphones, and speakers linked by satellite to the armchair traveler who remains at home. After renting a proxy in the city of your choice, you command it to explore the foreign streets and sights, watching through its eyes on your TV set and piloting its movements using a remote-control console. You walk your proxy toward the city's shopping area, taking photographs with a handheld Minolta thoughtfully provided by the local robot hire agency. The proxy arrives at the marketplace and you use it to pick out souvenirs, haggle with shopowners over the price, and have the purchases sent to your home, paying with a credit card you authorized for the robot.

While early models of proxies might provide only visual and auditory information, Yates speculates that later models would give their owners a complete sensory experience. You'll actually taste that frothy cappuccino from the cafe in Rome and feel that luxurious Japanese silk. The high cost of a proxy's telepresence equipment, Yates admits, will make it unaffordable to the average consumer for years. And even when the price comes down, specific safeguards must be built into the system to stop criminals from stealing the proxies and using them to mug the elderly, rob banks, plant terrorist bombs, or commit murder. Once perfected, however, proxies might provide a mind-broadening and safe alternative to travel "in much the same way," Yates says, "that cars provided a new and exciting alternative to walking."

If teleworkers can project their efforts thousands of miles across the face of the Earth, why not also just a few hundred miles up—into space? Because of the hostile environment and the tremendous expense of transporting people into orbit and maintaining or rescuing them there, NASA officials expect that most construction and routine maintenance activities will be performed either by robots or by teleoperators. Early concepts from the late 1960s included a manned space tug bristling with robot arms, which may have inspired the "pods" seen in the film *2001: A Space Odyssey*.

Probably the best-known space teleoperator is the Space Shuttle's robot arm—the Remote Manipulator System, or RMS. The RMS has a 50-foot reach and takes about half a minute to complete any motion. But why hurry? In zero gravity nothing weighs anything—you can use a 100-pound, long, slender pipe to move a 32,000-pound, 60-foot-long, 15-foot-diameter cylindrical load very slowly. The Shuttle will gain a second arm on the starboard side in future missions. Conceptually the RMS is much like a human arm, with yaw and pitch at the shoulder joint; pitch at the elbow; and yaw, pitch, and roll at the wrist. However, RMS does not have a real "hand" but only a primitive grappling fixture, so it can only grasp objects equipped with special knobs.

A number of concepts for free-flying remote space manipulators have been pursued, though none is funded as yet. Right now, the main space agency focus for teleoperation and robotics research is the Remote Orbital Servicing System (ROSS) concept, developed by engineers at NASA's Langley Research Center and Martin-Marietta's Advanced Automation Technology lab. ROSS is planned as an anthropomorphic teleoperator with a stereo camera and two arms as versatile as those of a space-suited astronaut. It could be built now with existing technology. ROSS would fit snugly inside the Shuttle cargo bay and could fly out from the Shuttle or from a space station hangar to service satellites. It would be operated from a ground control station with visual feedback and gripper controllers. Estimates are that a telepresence servicer as dexterous as an astronaut in a space suit could be designed, developed, and built for $50 million to $100 million.

Control is always a problem because of the time delay, a minimum of half a second and more likely one to two seconds, resulting from all the satellite and ground relay links the control signals must pass through. As a result, says Alfred Meintel, chief of Langley's Automation Technology Branch, "the operator normally would function more as a director or supervisor of the system." Supervisory control will initially be implemented for simple jobs, such as changing manipulator hands or moving a manipulator arm to a standard position. As confidence grows, more complex tasks will be entrusted to the robot. For example, a telepresence operator might tell the machine to *Remove Panel*. The supervisory control system decomposes this command into numerous low-level tasks—it finds the desired panel, figures out the proper manipulator motions to remove the panel, loosens the bolts with a screwdriver, then finally stows the removed panel.

The proposed American manned space station offers an excellent opportunity to develop truly sophisticated space telepresences. A humanoid robot could execute laboratory research tasks in the absence of the principal investigator, preparing slides for analysis, monitoring crystal growth experiments, and performing space station maintenance. Robots would need the same access to the station as a human, and these devices would either have to be exceedingly reliable or have the ability to fix each other. For really large jobs just outside the station, a tentacle manipulator might be used. Flexible teletentacles could be hundreds of feet or even miles in length, and could snake around into all sorts of inaccessible places.

One highly unusual space station teleoperator facility was proposed by a recent NASA Ames Research Center/Stanford University summer study group, which recommended a telepresence "guest chef" program. "Assuming that the ingredients

Zero-g cowboys rode this three-armed, rider-controlled "space horse,"
one of a series of experimental vehicles built for NASA.

necessary to prepare food in space are available aboard the station," reads the group's final report, "cooking is so time-consuming that one obvious solution might be to construct a versatile teleoperated or robotic chef." The robochef would retrieve ingredients, open containers, perform necessary measurements and blending operations, and control food placement and removal from the oven as well as recycling, restorage, and cleanup tasks. Great human chefs on Earth could whip up dishes literally "out of this world" by remotely operating station kitchen facilities on a guest chef basis, enabling astronauts to eat better and make long tours of duty literally more palatable.

The space station could also be used as a control center for wide-ranging space telepresence activities. Ground-based operators suffer nearly intolerable feedback delays, whereas station-based operators could service satellites *in situ* with essentially no time delay, enabling more delicate manipulation with full telepresence ability. A space-controlled telepresence Extra-Vehicular Activity (EVA) system is a realistic goal for the 1990s. It could potentially handle satellite servicing, remote spacecraft refueling, and long-duration work assignments in high-altitude geostationary orbits; it would be less expensive than manned EVA, and safer.

According to W. David Carrier, III, of Bromwell Engineering in Florida, we may be able to set up a teleoperated lunar mining facility with only one full-time person on the Moon to monitor operations for maintenance and troubleshooting. Explains Carrier: "It should be possible to remotely control front-end loaders from Earth. The loaders would be equipped with television cameras and various sensors to monitor their performance and location. This data would be displayed to an Earth-based technician who would control the operations." Carrier says the technology exists right now. "On a small scale, remote-controlled excavation on the Moon and Mars have [*sic*] already been acomplished in the Surveyor and Viking programs."

"The fact of a three-second time delay in the Earth-lunar control loop leads to instabilities," cautions Thomas Sheridan of MIT, "unless an operator waits three seconds after each of a series of incremental movements. This makes direct manual control time-consuming and impractical, so supervisory control is imperative." Indeed, psychological studies show that the feeling of telepresence fails when sensory feedback delays exceed one-tenth of a second, representing a maximum distance of about 10,000 miles at the speed of radio waves.

Teleoperation can also be used for space exploration. The Soviet Union deployed Lunokhod I, an unmanned rover, on the lunar surface in November 1970. It was operated totally by remote command from an Earth-based crew of five, who drove the vehicle using wheel-rotation commands to each of eight solar-powered electric motors (one per wheel). Pictures from a pair of TV cameras mounted at either end were transmitted back to Earth, and an inclinometer gave vertical attitude and acted as a safety device. Lunokhod I explored a path about a mile long and five hundred feet wide.

Clearly we can teleoperate Earth-orbiting or lunar devices, but beyond the distance of the Moon all bets are off. Time delays to Mars are on the order of a half-hour, and a signal to a spacecraft out near Pluto could take the better part of a day for one round-trip message. Another serious problem is the wide (thus expensive) communications bandwidths needed to handle TV and control signals for a complex remote device. While it's true that the Viking Mars landers, Voyager missions, and the forthcoming Jupiter Orbiter Probe employ a limited form of supervisory control, they are all largely automated. The same will hold true for future planetary rovers.

That doesn't completely rule out translunar telepresence, though. Astronauts orbiting a planet could teleoperate mobile manipulators maneuvering on the surface. A Mars rover with good telepresence manipulators could make extensive excavations, then configure scientific equipment under human control to exploit what has been discovered. With good enough equipment, space crewmen could even pilot remote-controlled aircraft through the atmosphere of alien worlds.

A bit further off in the future are the many possibilities for remote-controlled medical care and surgery. To date, telecommunications links have been used exclusively to observe, diagnose, and advise patients at distant locations, and for medical education purposes. One of the first telecommunications systems designed to support outpatient care in remote areas was initiated in 1971 at the Papago Indian Reservation near Tucson, Arizona, under NASA sponsorship. Patients visited a mobile van and a nurse operated the remote diagnostic equipment, permitting transmission of heartsound, slow-scan images of the patient, electrocardiograms, x-ray photos, and images viewed through microscopes and endoscopes. In another experiment in the mid-1970s, physicians at the Massachusetts General Hospital provided medical care to a thousand patients several miles away at the Logan International Airport Medical Station by means of a two-way audiovisual microwave circuit. The nurse used a telestethoscope and transmitted the patient's electrocardiogram, pulse rate, respirations, systolic blood pressure, and other physiological data to the remotely located doctor. A telemicroscope was employed to view blood smears and urine sediments, and x-ray photos were also transmitted and correctly interpreted using this system. Today, simpler but related equipment is widely used by paramedic teams in many major cities.

Practitioners of telemedicine tend to see themselves as pioneers, but actually they are pretty conservative. When asked about the future of telehealth, Canadian Health and Welfare Technology Consultant David L. Martin exclaimed: "There has *never* been any intention to provide remote-controlled medical services! At most, supervision of health care services has been provided." Martin cites a recent example in which surgeons from London, Ontario, supervised a surgical procedure at Moose Factory, some 2,000 miles away across the Canadian tundra, by means of two-way, satellite-relayed, closed-circuit television. "However," noted Martin, "it was a surgeon at Moose Factory who operated, with London providing advice." In a candid appraisal of his own profession, marine corps Colonel Paul W. Brown of Fitzsimons General Hospital in Denver admits that one of the biggest problems is dealing with the temperament of the surgeon. "A lot of us are prima donnas," he admits, "and we are not about to recognize that any machine is going to supplant our wonderful healing hands."

The best candidate for telesurgery is probably the eye. It is one of the most delicate organs of the human body, easily disrupted by even slight surgical interventions. It is easy to get at, being one of the few organs lying exposed to the external environment, and is readily immobilized for relatively long periods of time without harmful effect. Surgery is time-consuming and difficult, entire procedures taking place within a thimble-size area. This puts a great strain on surgeons, since it ties them to the operating microscope and virtually immobilizes them for hours. Understandably, their precision of movement decreases with increasing length of the procedure, yet the most delicate manipulations (those close to or in direct contact with the retina) usually occur in the late phases of the operation. Given these facts, it was inevitable that the first attempt to build full-scale telesurgical systems has been by ophthalmologists.

The first crude micromanipulator for eye surgery was designed and tested in 1956 by Dr. C. E. T. Krakau at the Eye Clinic, University of Lund, in Sweden. It used electrical servomotors to achieve smooth, nonmanual movements within the eye, and was tested on rabbit eyes with limited success. In 1972, researchers at the University of Illinois and at the UCLA School of Medicine independently came up with two distinct micromanipulator systems for intravitreal (within the eyeball) surgery, but both were entirely manual with only limited utility. By 1979, the UCLA device, developed and tested jointly in over two hundred surgical procedures with the Jules Stein Eye Institute under the leadership of Dr. Manfred Spitznas, had evolved into the first true surgical teleoperator.

The base of the micromanipulator connects to the operating table. In the center of the base is a molded rubber headrest with skull clamps to hold the patient's head fixed during surgery. Attached to the base plate is an articulated arm holding a special ring that is sutured to the eye muscle to achieve complete immobilization of the eyeball. The actual micromanipulator unit is made up of three basic components: the straight micromanipulator arm, with a pivoting bar at the end; the micromanipulator arc, which is attached to the pivoting bar; and a receptacle cartridge to hold microtools, which rides on the micromanipulator arc. Left-right or forward-backward motion of the joystick produces similar movements of the intraocular instrument, while twisting the joystick moves the microtools up or down at varying rates of speed. The surgeon can command motions so fine that even under high microscopic magnification they are barely perceived. There is no force reflection from the microscalpel, but the surgeon views his work as a three-dimensional image in an operating stereoscope, which gives him a limited telepresence effect.

Spitznas points out that ocular telesurgery is actually superior to freehand methods, for many reasons—it allows work on an immobilized eye, eliminates surgeon hand tremor throughout the longest operations (up to eight hours), and offers greater comfort for the surgeon, uniformly high precision, stable instrument position, safe training of beginners, and the utmost accuracy even by less-experienced surgeons. The teleguided motorized micromanipulator represents the first fully successful attempt to separate the hand of the surgeon from the surgical instrument to obtain a higher degree of accuracy. Says Spitznas: "The device adds a new dimension to surgery, opening the door to the future development of still finer procedures that cannot be performed at all by the free hand."

Currently, the control unit is linked to the micromanipulator by wire. "I've thought of making a wireless [radio] connection, which should be rather easy to do," Spitznas recently told me. "This will become of great interest as soon as true stereo television is available." Once remote viewing facilities are added to the operating scope, the eye surgeon can be anywhere—the next room, the next city, or thousands of miles away—and still perform the surgery successfully after a local practitioner has properly prepared the patient.

A limited form of supervisory-controlled telesurgery has been achieved using the Model 771 Microscan micromanipulator manufactured by Laser Industries, Ltd. in Tel Aviv. Last year, surgeons at the Department of Neurosurgery of the University of Tennessee used the micromanipulator to remove a tumor by carbon dioxide laser surgery. The area of tissue to be ablated is selected by the surgeon and outlined using a helium-neon aiming beam and the joystick on a remote-control panel. The coordinates of this area are stored in microprocessor memory and may be verified by having the device retrace the selected area. If the surgeon is satisfied he proceeds with the laser operation. If not, the process is repeated until a satisfactory tracing is obtained. Microprocessor control increases the precision of the laser dissection and decreases the fatigue experienced by the surgeon by eliminating the repetitive movements of manual micromanipulators.

Why have these techniques not been more widely applied in microsurgery? Microsurgery has four general applications: replantation of severed parts; tissue transfer from one part of the body to another; reconstruction of severed nerves in hands, arms, and legs; and operations such as reopening fallopian tubes and reversing vasectomies. Probably one major reason there are still only a few teleoperated instruments is that no telemanipulator yet exists with the required dexterity, precision, and reliability needed to sew capillaries no thicker than narrow string and nerve fibers the size of fine thread.

Another problem to be overcome is acceptance by surgeons. Most probably would not trust a telesurgical system unless its reliability was extraordinarily high. Interruption of sensory feedback or control for even a brief moment could be fatal if not backed up by fail-safe or supervisory control mechanisms. "If a specialist is operating remotely and something goes wrong," observes David Martin, "you can be assured he'll want someone there he can trust to be his eyes, ears, and hands, and to do precisely what he requests—to 'pick up the pieces,' so to speak." Today, most surgeons would likely refuse a telesurgical system even if it existed, insisting instead that the remote patient be transported back to them. But this might not be feasible in space, in a deep-sea habitat, or at an Antarctic base—settings that may motivate the widespread use of telesurgical technology in the not-so-distant future.

If we can miniaturize teleoperators, there are clear and urgent applications in surgery and biology. For instance, surgeons today can repair and replace small blood vessels, but not in the brain and other organs where scalpel and forceps cannot reach. For that, the microvascular telesurgeon will need tough, force-reflecting microhands on slender probes that can reach through very narrow passages. The U.S. annual expenditure for coronary vessel repair already exceeds $1 billion, so there is big money involved that could be used to develop sophisticated microtelepresences. A small repair to the brain of a stroke victim could yield an extra decade for an intellect that took a lifetime to build.

Today, especially in Japan, stones and emboli are removed and pacemakers implanted with simple, remote-controlled probes. These limit repairs to a narrow spectrum of cutting, crushing, and stretching operations. Telepresence minihands would permit surgeons to attempt procedures that seem impossible today. For example, microsurgeons today can replace two or three major coronary vessels, but not a hundred smaller ones. A telemicrosurgeon could direct a semi-intelligent procedure to swiftly make numerous small repairs. The device would interrupt to ask about unexpected difficulties.

What about incisionless surgery? One possibility is the concept of remote-controlled "medical mites" made feasible by modern micromachinery technology. Already, using techniques similar to those employed in silicon chip fabrication, engineers have constructed microthermometers and pressure sensors, microrefrigerators, a micro-gas-chromatograph and a micro-ink-jet nozzle. A millimeter-size silicon cantilever beam was fabricated for a tiny microaccelerometer small enough "to be sutured to the heart to measure its acceleration," according to Lynn M. Roylance of Stanford University, who built the device. Some medical mites would be like microminiature submarines, released inside the human body for internal sensing. Other mites could float, crawl, or swim through major arteries in the human body and perform on-site repairs from within, controlled by radio link under direction of a skilled telemicrosurgeon.

Telesurgery will also make it possible for a doctor in one corner of the world to operate on a patient anywhere else. Such a system will require electronically slaving a set of sophisticated mechanical manipulators at the remote site to the arms of the physician so that he can literally reach around the world, if necessary. Whole teams of doctors, each in a different country, could join together via telesurgical links to operate on a difficult case. In the remote operating theater of the future, surgeons no longer will scrub up because their hands never touch the patient. Patients are anesthetized by remote control, prepared for surgery by robotic nurses, and operated on by specialized robotic surgical tools, all teleoperated from somewhere else. The operating room remains perfectly sterile, scalpel movements are extraordinarily precise, and fatigue is no longer a major factor. Supervisory control will permit telesurgeons to issue high-level commands to their robot stand-ins, such as, "Suture this artery," or "Five-percent IV glucose stat."

Symbiotic bionics, or symbionics, has found perhaps its broadest expression in the mechanical assistance of handicapped persons. These systems include mobility aids, such as wheelchairs and walkers; remote telemanipulators or robot nurses; powered orthotic braces, which fit over paralyzed limbs and force them to move under machine power; and prostheses, artificial organs, and synthetic limbs that actually replace a missing appendage.

For example, a voice-controlled wheelchair-manipulator was developed in the late 1970s by NASA's Jet Propulsion Laboratory. This device can pick up packages, open doors, turn a TV knob, and perform a variety of other functions. The patient teaches the wheelchair to recognize thirty-five one-word commands (e.g., go, stop, up, down, right, left, forward, backward) by repeating them a number of times. After training, the computer analyzer responds only to the patient's voice, translating commands into electrical signals that activate appropriate motors and cause the desired motion of chair or manipulator. The wheelchair-manipulator has been tested at Rancho Los Amigos Hospital in Downey, California, and is being evaluated at the Veterans Administration Prosthetics Center in New York City.

A voice-activated passenger car has been developed by the Nissan Motor Co. of Tokyo, Japan, for drivers who have lost the use of their hands. Apart from the steering, braking, and acceleration systems, a computer activates the car's mechanisms through vocal command. At the command of the driver, the computer controls lights, windshield wipers, windows, mirrors, and even the position of the driver's seat. The car starts on voice command, and the driver controls the car's direction and speed through specially designed foot pedals. Nissan has not announced plans for export, but says an ordinary car could be fitted with the computer system for about $4,000.

Disabled people can now also be provided with a remote-controlled telemanipulator or robot nurse. At the Veterans' Administration Medical Center in Palo Alto, California, engineer Larry Leifer of Stanford University has customized an industrial robot to be an electronic servant for the handicapped. Leifer's robot nurse has a 3-foot-long robotic arm with a slender mechanical grasper and an omnidirectional wheel base to allow movement around a room. It speaks and understands about seventy words. Designers deliberately avoided using an independent, intelligent robot, opting instead for a system controlled by people.

As with the JPL wheelchair, the operator talks the machine through motions to perform tasks as complicated as cooking a simple meal, serving the food, salting it, serving beverages, typing, playing board games, turning the pages of a book, picking up a telephone receiver, punching the keys on a pocket calculator, or giving a shave. Already ninety patients have trained with Leifer's robot, and within a decade a mass-produced version should be available for the price of an average car. "In the next ten years," says Leifer, exuding confidence in his Robotic Aid Project, "robots functioning as nurses and domestic servants will be as common as cars." They will be widely employed, he says, for useful household work in nursing homes, to care for the elderly, and in hospitals for routine work and intensive care.

Japanese engineers at the Mechanical Engineering Laboratory of the Ministry of International Trade and Industry have devised a robot guide dog for the blind, Meldog Mark I. The prototype robot walks 3 feet in front of its master, matching its speed to his by an ultrasonic ranging sensor. The master moves in a preset safety zone. If he strays from the zone, the robot warns him, using electrocutaneous stimulation. Future versions will be able to detect obstacles or dangerous situations ahead and navigate the streets using a stored, computerized map of the city. The Japanese dogbot is interesting because one of its projected functions is "intelligent disobedience" of the master when danger is detected, for the master's own safety—possibly the first planned implementation of Asimov's famed Three Laws of Robotics (see page 49).

Another approach for assisting the disabled is the orthotic brace. Probably the most ambitious of these is the hydraulic-powered, computer-controlled exoskeletal walker for paraplegics devised by Ali Seireg and Jack Grundman of the University of Wisconsin. The device performs such tasks as walking, rising from a seated posture, sitting down, stepping over obstacles, and climbing stairs for paralyzed patients. These actions are preprogrammed motion sequences and can be executed at different speeds and in partial or full cycles. The Wisconsin Walker weighs 75 pounds and easily supports a 165-pound operator. However, it cannot balance on its own, so patients need two canes to steady themselves. Seven control switches are mounted on one of the canes within easy reach. Similar walkers have been developed by teams at the Institute for Automation and Tele-communications in Belgrade, Yugoslavia, and at Waseda University in Tokyo, Japan.

Prosthetic devices that replace a missing limb provide perhaps the purest form of telepresence. Modern prostheses tap the myo-electric currents generated by a muscle fiber when it contracts. Signals from brain to the muscle fiber, in, say, an arm stump, are picked up by electrodes, amplified by tiny motors in the prosthetic device, then used to control the action of the device in an effortless, wholly natural way. One commercially avail-able limb appliance, the Utah arm, was developed by Stephen Jacobsen and his colleagues of the Center for Biomedical Design at the University of Utah. The Utah arm flexes at the elbow, rotates at the wrist, and manipulates fingerlike attachments that can hold forks, bottles, or pencils. Using sensors on a sub-ject's arm stump, tiny muscular contractions are interpreted by a computer and control the delicate motions of the artificial limb. Frank Clippinger, Jr., of Duke University Medical Center made an arm with electrocutaneous feedback that gives a simulated sense of feeling and touch that patients easily learn to recognize. Vert Mooney at Rancho Los Amigos Hospital in Downey, California, built a prosthetic arm that gave its first wearer, a karate expert, a grip strength twice that of a normal man.

There are artificial legs, too. The best of these is the result of a joint project of Moss Rehabilitation Hospital and Drexel and Temple Universities. The Drexel-Moss artificial leg uses subcon-scious and conscious signals sent to the above-knee muscles used for walking. Rehabilitation engineers attached nine electrodes to the patient's thigh and had him think of taking a step, think of extending his leg, and so forth. Each time, a computer re-corded and classified the gross electrical activity in his muscles. Later, these distinct activity patterns were translated into spe-cific instructions to a pneumatic, piston-driven artificial leg. The big advantage of the Drexel-Moss system over its predecessors is its naturalness—it allows automatic walking but can also make conscious maneuvers, such as obstacle avoidance and stumble recovery, as well. Right now the leg is tethered to a big computer in the lab, so engineers are working on a self-contained microprocessor so the patient can walk out of the lab under his own power.

A wide variety of symbionic organs are now available, too. Remotely programmable and rechargeable pacemakers can be purchased, and simpler heart prostheses have been installed in hundreds of thousands of patients worldwide. Medtronic, the largest manufacturer of pacemakers, is bringing out an artificial pancreas that dispenses insulin using an implanted microchip-controlled pump. The device has radio telemetry that enables remote physicians to interrogate your organ to find out how much insulin remains in the reservoir or to reprogram the medication flow rate. (The diabetic can also directly control his own bionic organ using a hand-held unit.) Researchers at the University of Minnesota are working on a biotech device that will control and supplement the flow of neurotransmitters in the brain, a potential boon for those who suffer from Parkinson's disease and schizophrenia.

In *Foundation's Edge*, Isaac Asimov's far-in-the-future space hero commands his ship's computer with his mind. A direct linkage between human minds and the robots they control is the ultimate result of the man-machine symbiosis. Is it possible that we may someday be able to control a robot from a distance, using our brains to "think" its every move?

Thought control, or biocybernetics, has been seriously studied by the military since the early 1970s with the goal of having a pilot control an aircraft and weapon systems by thought. Sensors picking up the brain's electrical activity produce signals that are processed by pattern recognition equipment, activating any system as though a switch had been pressed. According to Gene Adam, chief of the advanced crew station branch at McDonnell Douglas, operational use of this technology is fifteen to twenty years in the future, although applications could be available within the next ten years for simulator studies.

Biocybernetic programs are being pursued at numerous universities and think tanks including the University of Illinois, MIT, UCLA, the University of Rochester, and Stanford. Neurophysiologist and electrical engineer Lawrence Pinneo of SRI

International in Palo Alto, California, has made a "thinking cap" that picks up a subject's electrical brain wave activity via scalp electrodes, analyzes them through a computer, and then translates them into action. Pinneo's volunteers can move dots from side to side on a computerized TV screen; they have even been able to "think" an object through a video maze. Pinneo's executive computer can also recognize words, spoken aloud or silently thought, by comparing them to prerecorded characteristic brain wave patterns of the particular subject. Erich Sutter, a biomedical engineer at the Smith-Kettlewell Institute in San Francisco, has built and tested a practical system using EEG electrodes and a video screen with flashing symbols to operate home appliances remotely. It will probably be put on the market in a few years for as little as $5,000.

Symbiosis implies partnership and a sharing of responsibilities. Just as we are learning to remotely control machines with our brain waves, computers are slowly gaining the ability to control our bodies. In the simplest case, a patient's implanted artificial pancreas is commanded by a remote physician to double the daily dosage of insulin. But computers can also control our limbs and brains to a limited extent, using direct electronic stimulation of the brain (ESB). ESB refers to the implantation of tiny electrodes deep within the living brain. These electrodes are pulsed with minute quantities of electrical current that interferes with the normal processing of signals, resulting in altered motor or behavioral patterns.

Extensive remote motor control has been demonstrated in animals using implanted brain electrodes. Lawrence Pinneo and his team at SRI once implanted a dozen electrodes in the brainstem at the back of a monkey's head. Small portions of the animal's motor cortex had been surgically disconnected for the experiment. Pinneo's device, the Programmed Brain Stimulator, fired the electrodes in the proper sequences to evoke motor responses from the monkey. One programmed sequence, for example, caused the animal to reach out with its paralyzed arm, grab a piece of food, and return this to its mouth—all under computer control. Another sequence had the monkey reaching around to scratch its back, a complicated series of arm and wrist motions. The motor cortex was mapped in more than two hundred locations. Experimenters learned exactly which parts of the brain controlled wrist flexion, knee- and hip-twisting, and grasping movements. It appears that extensive motor telecontrol is possible. In a series of tests on laboratory cats with suitable electrode implants, reported one researcher, feline eye pupils were controlled "as if they were the diaphragms of cameras."

In man, cerebellar pacemakers, an ESB technology, are employed today to treat serious pathological behaviors. Electrodes placed in the pleasure centers of the brain are stimulated to help restore normal mental functions. Over the years, ESB has been used experimentally on human subjects to evoke pain and ecstasy, fear and friendliness, and "cooperative attitudes" in previously recalcitrant patients.

Is it possible to conceive of teleoperating *people*? Jerrold Petrofsky has already teleoperated a laboratory cat! In preliminary experiments a few years ago, the scientist connected myoelectric sensors to his own legs, with numerous wires trailing off to a computer, then more wires leading to the nerves of a nearby paralyzed cat. With the system activated, Petrofsky straightened his right knee and the cat's right hind leg shot out. He flexed his leg, and the cat's leg bent in unison. For several hours Petrofsky and the wired cat kicked their legs about in perfect synchrony.

To teleoperate a person, we need both motor and sensory control. Given plenty of additional research to precisely map the human motor cortex and a large enough number of electrode brain implants, it seems likely that neurotechnicians could make a wired person hop around like a zombie. What about remote sensing through another person's eyes or ears? To achieve telepresence with another living being will require sophisticated optic and auditory nerve taps which lie many decades beyond current technology. Scientists are only beginning to map the geography of the human mind—its microneurography—using techniques of nerve traffic analysis. There is far more going on than we yet understand. Still, one wonders what it would be like to telepresence a dog or cat? A bird or a dolphin? A friend or lover? An enemy?

In 1976, Adam Reed, a postdoctoral psychologist at Rockefeller University in New York City, took an even further look when he claimed that within fifty years miniaturized computers implanted under the scalp will be programmed to read and speak the electrochemical language of the human brain. Once this is accomplished, information can be fed directly into the brain's data processing units without going through peripheral equipment such as eyes and ears. No need to read a book—the computer just squirts its contents into your head. To achieve these results, Reed estimates that at least 100,000 electrodes per square millimeter will be required in the implanted matrix.

If and when this technological feat is achieved, the symbiosis of man and machine will be virtually complete. As people communicate directly with computers, they will gain instant access to the databanks, data processing capabilities, and telecommunications networks of the entire human race. Merely thinking of someone you wish to talk to might initiate a search by your computer symbiote to locate that person anywhere in the world and establish direct contact. Thoughts would flow between beings—human or animal—in seemingly telepathic fashion.

Glenn F. Cartwright, director of the Division of Educational Computer Applications at McGill University in Toronto, calls these intelligence amplifiers "symbionic minds." Symbionic minds may signal a new and different relationship among the peoples of the Earth and represent the beginnings of true global consciousness. "The new symbionic mind will act purposefully and willfully," says Cartwright, "but always on our behalf and at our direction. It will be our constant companion and friend, our conscience and alter ego. The symbionic mind will mark the next step in the evolution of humankind to a higher plane of existence and the dawn of a new era."

The Ultimate Worker
Joseph F. Engelberger

We may expect robots to go where men fear
to tread. They may live in poisonous atmospheres and at temperatures intolerable to man.
. . . It is conceivable they will be more than the
ultimate workers, they may be the ultimate survivors should humans become so intemperate
as to destroy our Earth's life support system.

If robots could write, the Human Resources Manager of a large
factory might see the following form slide across his desk:

APPLICATION FOR EMPLOYMENT

NAME: Unimate 2000B   SOCIAL SECURITY NO.: None

ADDRESS: Shelter Rock Lane, Danbury, CT 06810

AGE: 300 hours (by clone extension, 16 million hours)

SEX: None   HEIGHT: 5 ft   WEIGHT: 2,800 lbs

LIFE EXPECTANCY: 40,000 working hours (20 man-shift years)

DEPENDENTS: Human employees of Unimation Inc.

NOTIFY IN EMERGENCY: Service Manager, Unimation Inc.

PHYSICAL LIMITATIONS: Deaf, dumb, blind, no tactile sense,
one-armed, immobile

SPECIAL QUALIFICATIONS: Strong (100-lb. load), untiring
(24 hours per day), learn fast, never forget except on command, no wage increase demands, accurate to 0.05" throughout
spheres of influence, equable despite abuse

HISTORY OF ACCIDENTS OR SERIOUS ILLNESS: lost hand
(since replaced), lost memory (restored by cassette), hemor-
rhaged (sutured, and fluid replaced)

The title of this chapter—"The Ultimate Worker"—implies that some sort of robotic technological forecast is forthcoming. And so it is—we'll be taking a look at the *ultimate* ultimate worker that robots will soon become. But before we do, we should note that the robot "oafs" that now exist are already doing a lot of jobs, doing them well, doing them to their owners' economic advantage, and are still awaiting more opportunities.

The first industrial robot went to work over twenty years ago, in 1961. Its job was to stand at a die-casting machine, extract the hot metal castings, and dunk them in a warm bath. Technology has advanced since then, but the 1961 machine is perfectly capable of doing now what it did then. In fact, many of the robots of the 1960s have logged over 100,000 on-the-job hours and are still going strong. And considering that 100,000 work hours are roughly equivalent to 50 man-years, you can conclude that these old robots are no more susceptible to being retired than the human who is putting in his/her years before retirement.

To continue with the analogy, there is the added fact that these old-timer robots cannot be easily retrained to do more sophisticated tasks. They could not, for example, sense if they were welding in the right place, nor could one pick up a part off a moving conveyor belt. But until present die-casting processes change dramatically, or become obsolete, the robots of the 1960s will continue to earn their keep.

At Unimation Inc., we used to report each year how many hours all of our robots had logged, but after the machines had put in about 15 million hours, we stopped adding up the hours— the figures were no longer meaningful. What really counts is how many jobs robots have done for sufficiently long periods of time to prove themselves. And since robot skills are communicable, we may presume that the new machines can be imbued upon delivery with the same skills to carry out the same factory jobs with the same ability as the first robots.

Manufacturers who want to compete in the world marketplace would do well to appreciate just how much can be accomplished by deaf, dumb, and blind robots in industrial jobs whose demands are surely subhuman. Robots already exist to do these jobs, and they are technically and economically superior to their

A worker robot and his human teacher.
Besides simple jobs, steel-collar workers are expected to learn
more difficult types of welding.

human forebears. There's no need to sit back and wait, claiming that smarter robots are on the horizon. Of course they are, but many of the jobs being done today do not need capabilities beyond what contemporary robots possess. In a way, waiting for the smarter machine is a little like hunting for the ultimate die-casting operator by waiting for an astrophysicist to apply. And yet, a lot has happened in twenty years; robots can already do much more than simple die-casting work.

But first let's get the basics straight and answer some simple questions. First: *What is a robot?* As might be expected, Webster has a definition: "an automatic apparatus or device that performs functions ordinarily ascribed to human beings, or operates with what appears to be almost human intelligence." This definition is pleasing to a roboticist because it links the robot concept to its science-fiction origins and leaves the future open to unlimited possibilities from research and development.

Ah, but Webster is not enough for the press or the public. They push for a definition that's more exciting, and they're completely dismayed when faced with the definition of an industrial robot concocted by the Robotics Industries Association: "A robot is a reprogrammable multifunctional manipulator designed to move material, parts, tools, or specialized devices through variable programmed motions for the performance of a variety of tasks."

When all else fails I throw up my hands in dismay and say, "I cannot define robot, but I know one when I see one." After all, human beings come in a great variety of sizes, shapes, colors, and sexes with a great range of skills and a broad range of intellectual prowess. Yet it's easy to recognize a human when one sees a human.

There exists a range of different kinds of industrial robots busily at work in factories around the world, and they too are recognizable. There is always an arm, and the arm can reach into its environment to grasp materials or position tools. Most are fixed at their workstations, because moving them from job to job is an exercise of some complexity. Most U.S. and European manufacturers presume a robot has many articulations, all of which are servo-powered and computer-controlled to allow the robot arm to be placed with infinite adjustability anywhere in its sphere of influence.

One steel arm places a tire and wheel on a new car.
Automation has generally been credited
with helping to save the U.S. auto industry.

The Japanese, who are now the most advanced in robotics, are taking a different tack and have greatly broadened their definition of industrial robots. Their classifications range from the simplest arms to what they call "intelligent robots." Following is a list and description of this variety of machines:

MANUAL MANIPULATOR—a manipulator worked by a human operator.

FIXED SEQUENCE ROBOT—a manipulator that performs successive steps of a given operation repetitively according to a predetermined sequence, condition, and position. Its set information cannot be easily changed.

VARIABLE SEQUENCE ROBOT—a manipulator similar to the fixed sequence robot, but whose set information can be changed easily.

PLAYBACK ROBOT—a manipulator that can reproduce operations originally executed under human control. A human operator initially operates the robot to feed in the instructions—relating to sequence of movement, conditions, and positions—which are then stored in the memory.

NC (NUMERICALLY CONTROLLED) ROBOT—a manipulator that can perform a given task according to the sequence, conditions, and positions commanded via numerical data, using punched tapes, cards, or digital switches.

INTELLIGENT ROBOT—a robot that can itself detect changes in the work environment, using sensory perception (visual and/or tactile), and then, using its decision-making capability, can proceed with the appropriate operations.

The robot industry is not very well organized anywhere in the world, but the Robotics Industries Association in interviews around the world made its global population estimates of industrial robots as shown on page 191.

A true feeling for how an insensate robot may take over for a human worker will become apparent in the following vignettes, the answer to my next basic question: *What can robots do?*

	TYPE A	TYPE B	TYPE C	TYPE D	TYPE E	TOTAL
JAPAN	28,900		17,500	7,347	53,189	67,435
UNITED STATES (USA)	400	2,000	1,700	600	40,000	44,700
WEST GERMANY	290	830	200	100	10,000	11,420
SOVIET UNION (USSR)						3,000*
SWITZERLAND	10	40			8,000	8,050
CZECHOSLOVAKIA	150	50	100	30	200	530
GREAT BRITAIN	356	223	54	80		713
POLAND	60	115	15	50	120	360
DENMARK	11	25	30	0	110	176
FINLAND	35	16	43	22	51	167
BELGIUM	22	20	0	0	82	124
NETHERLANDS	48	3	5	0	15	71
YUGOSLAVIA	2	3	5	0	15	25
SWEDEN	250	150	250	50	100	800
NORWAY	20	50	120	20	50	260
FRANCE	120	500			38,000	38,620
AUSTRALIA			62		120	182
ITALY						753 **
TOTAL	1,774	10,924	2,584	8,299	150,052	177,386

TYPE A. Programmable, servo-controlled, continuous path
TYPE B. Programmable, servo-controlled, point-to-point
TYPE C. Programmable, non-servo robots for general purpose
TYPE D. Programmable, non-servo robots for die casting and molding machines
TYPE E. Mechanical Transfer Devices (pick and place)

*Figures on the Soviet Union were supplied by Daiwa Securities, New York, NY.
**Italian figures based on 1980 completed survey.

Die-Casting Die-casting was the very first job entrusted to an industrial robot. This task was a likely choice because the job of standing in front of a die-casting machine and extracting hot metal castings every 15 to 20 seconds is indeed a harsh and brutal one. The working area is of course hot; the chemicals used to release the casting from the die release fumes; and an unwary die-casting machine operator who stands too close to the machinery might well be sprayed with molten metal. It's also a good job for a robot because the casting pickup position is predictably fixed. The metal starts as goo and hardens in a fixed location that has been taught to the robot. When the die opens, the robot arm can reach into the die bed, grasp the hot casting by the edge, and maneuver the casting from the casting machine.

At first all the robot did was extract the casting and then quench it in a water bath. But as robot users became more confident in their "employees," the jobs given robots at the die-casting machine workstation were more complex. Some robots not only extract the casting and quench it, but also transfer the casting to a trim press for removal of flash, or excess material.

In other instances the robot may be asked to place special inserts in the open dies before they close. In other complex operations the robot turned out to be great at spray-cleaning and lubricating dies as well. Because closing the dies for a subsequent shot when leaving cold metal in the die could result in die damage, some installations effectively provide the robot with very rudimentary sensory perception by switch or infrared signals to warn if the complete casting has not been extracted.

Spot Welding Spot welding—usually auto bodies—is the single, most ubiquitous application for industrial robots throughout the world. In this application the robot carries a welding gun, a tool that was traditionally wielded by humans on an automotive body assembly line. This job is a natural for a robot. The welding guns can weigh anywhere from 10 to 70 pounds, and manipulating them can be very tiring. A weary human welder could not be expected to place spot welds with the same dependability and accuracy as a robot worker. And because body assembly lines usually run on a multishift basis, the economic return to the manufacturer is quite high. The robot is fixed in a workstation and maneuvers the spot welding gun around the auto

Robots can do the same thing over and over with unflagging attention.
This machine applies more than 3,100 spot welds per car.

body (welding door frames, window apertures, rocker panels, etc.), placing a spot weld on the average of every one and a half seconds. A modern car body may have 1,400 to 1,800 spot welds placed by robots. The technology is well advanced, and production lines of up to 60 robots are not unusual. The robots may be mounted on both sides of an indexing line or even overhead.

Arc Welding Arc welding is much more broadly used in industry than spot welding, and the potential for robot application is correspondingly higher. However, to do any of this work the robot must have more sophistication than is necessary for spot welding. Primarily, the machine has to be able to move the welding tool over a continuous path at a constant speed. The spot welding robot need only hop from point to point and doesn't have to keep such a tight control on the tool path. Arc welding is an onerous job, with irritating fumes from welding materials, flying molten metal, and ultraviolet light that can be harmful to vision unless welding masks are worn. On the job, the sightless robot logs a much higher percentage of weld time than does the human who must stop continually to check his work. At today's level of robot technology, the arc welding done by robots is generally limited to high-volume runs where parts fit closely together and where tooling provides consistent fit-ups. This, unfortunately, is only a small percentage of all arc welding. In most instances tight tolerances are not maintained and the quality of the job depends upon the skill of the operator, including his visually lining up where the weld "bead" must be laid. Arc welding must be looked upon as fine here-and-now application for the blind robot, but one that offers a much greater opportunity to the sensing robot we'll be talking about later.

Glass Handling The most ideal handling jobs for robots are in industries that work with hot glass, such as making television picture tubes. Television funnels and face plates are formed from molten glass, and once a melt is started the operation has to proceed around the clock; after any shutdown it takes many busy hours to bring operation back into production. A robot that cannot deliver at least 99.7 percent efficiency in a hot glass fabricating facility is inadequate to the job, because lost time is so terribly expensive. Thus, robots that are unloading molds and placing parts in tempering furnaces have to work under extreme temperature conditions twenty-four hours a day, seven days a week, resulting in very high paybacks. This is particularly demanding upon the robot design because high reliability is essential.

Heat Treatment Heat treatment is a common metalworking activity, and once again an unpleasant one for a human operator and ideal for a robot. In addition to loading and unloading parts into furnaces, the robot may use its arm to position a torch to treat only specific areas of the parts.

Forging The horrors of the Industrial Revolution are often depicted by a forging plant where dirty, sweating humans shove the hot billets through the forging dies or under the forging hammers. This is desperately subhuman work and, sad to say, it is quite often too demanding even for today's robots, although they can stand the heat and carry the hot billets and endure the shocks and the noise with ease. But when billets come from a furnace in disoriented fashion, when they stick in forging dies, or are not at optimum temperatures, the blind hulk of a robot is at a loss.

Paint-Spraying This is one of the best jobs for robots. The robot must be able to follow a continuous path, but because of the sweeping spray pattern, accuracy is not critical. In the most common paint-spray robot control mode, an expert painter literally leads the robot arm around with the spray gun in operation. If the manually led robot sprayer has done a good job, the program is stored and then it can repeat the identical action. Paint-spray robots are now used worldwide wherever paint or sealer is being applied in long series or medium-length series production applications.

A robot paint-sprayer, of course, saves labor, but often the ancillary benefits are of even greater importance. A robot doesn't need to breathe and can stand high temperatures that may be useful to the work process but intolerable to human operators.

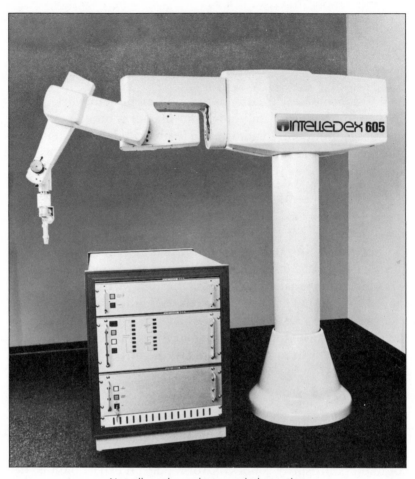

Not all worker robots are behemoths.
This robot comes in scales that vary from the size of a man
to one that fits on a tabletop.

Investment Casting This is really no more than a modern-day version of the ancient lost wax process that sculptors have used throughout history to cast works in bronze. In investment casting, a very accurate master mold is used to create a wax reproduction. Then the wax is alternately dipped in slurry and ceramic powder until a shell is built up. The shell is then baked hard; the wax melts, leaving a ceramic mold. The final step is to pour molten metal into the mold. The result is an extremely accurate part that usually requires no further finishing.

When Unimation engineers first looked at this process, they threw up their hands in dismay, because, in making the first dip into the slurry and in maneuvering the part to get an even coat, there was the problem of eliminating air bubbles on the surface of the wax. Just as we were about to give up, a clever development engineer remembered that, unlike humans, a robot can have continuously rotating joints. The solution was lovely. After the wax is dipped and worked to get a reasonably smooth coating of slurry and ceramic powder, it is spun at high speed. This throws off excess fluid and at the same time eliminates all surface air bubbles. The result is a perfect first coat and a very high yield of good parts from the roboticized workstation.

Robots breathed new vigor into this archaic manufacturing process. Shells of surpassing quality have been built with the robot as the primary piece of machinery in the system. Since some parts are large and result in shells that can weigh well over fifty pounds, the manipulation of the wax through the entire buildup of the ceramic shell becomes extremely tedious for a human operator. Investment casting is one activity in which a robot worker can constantly outclass its human counterpart.

Conveyor Transfer At first thought, transferring parts from one conveyor belt to another appears to be work that would hardly be challenging to a moron and therefore an easy task for a simple robot. Upon more careful consideration, it becomes apparent that a moron's abilities should not be underestimated. When parts are being transferred from one conveyor to another, our robot moron must sense when there are parts on the input conveyor and what the speed of that conveyor is. On the output conveyor, the moron must sense whether or not a section of it is already occupied. If there is an extended mismatch between

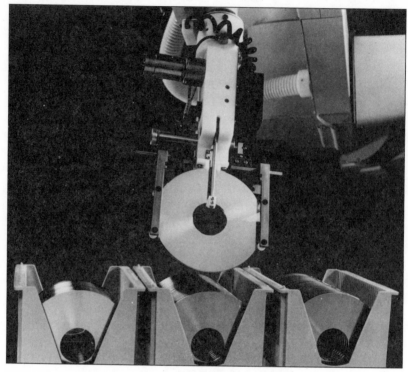

Not all materials-handling is of the weight-lifter variety.
This robot sorts delicate hard disks as they are rated by three machines.

the flow of input and the output conveyor, the operator will probably have to walk back and forth and match the input flow against the output flow.

The stationary robot that is assigned this "moronic" job must also synchronize its arm speed with the conveyors in order to remove and replace parts gently. If there is a mismatch and the robot finds itself removing parts from the input conveyor and having no empty hooks on the output conveyor, it must have the sense to place these excess parts in some kind of buffer storage location. Then, when the output conveyor has room for them, the robot must remember where the parts are stored and retrieve them to fill the output conveyor. The robot has to be pretty damn capable to carry all of this off.

The job is moronic, yes, but not too many of today's robots can cope with this conveyor transfer application. So there are still thousands of workstations throughout American and European industry where conveyor transfer is the sole activity of a human with an intellect very much superior to that of a moron.

Inspection The robot seldom does the actual inspection. Usually, the robot manipulates the parts and assemblies to be inspected into fixtures and, when inspection is completed, takes signals from the inspection station to direct the inspected parts to their destination.

While the operation of peripheral equipment is the most common application of roboticized inspection, it would also be practical to use the visual and tactile-sensing incorporated in the robot to allow inspection in a manner akin to that of the human who can see and feel. Thus, a robot with a sense of touch could reject a part that does not properly mate in its subassembly, and another robot with a vision module in its palm might reject a part whose finish was out of tolerance.

There are situations where robot inspection has an edge over human inspection, where it can use the absolute accuracy with which it can position its end effector in space. Thus, a robot carrying an instrumented wand could feel for the characteristics of an automobile body, such as location of wheel wells and door apertures, to ascertain that the body is within or outside of dimensional tolerance.

Economic Justification

Nobody *needs* a robot. Yep, that is absolutely true—even though the realization is a traumatic shock for a dedicated roboticist. The fact is that, at least in industry, there is little a robot can do that a willing human cannot do better. We struggle to justify the robot, we talk of hazard to human life, and we talk of the stultifying effect of rote activity on a human worker. But, realistically, these are "Goody-Two-Shoes" arguments that have little impact on a manufacturer who is devoted to maintaining profits in the face of keen competition. Even the intangibles like greater throughput and higher quality make a limited impact on manufacturing decision-makers. The crux of the matter is raw economics. A robot must justify itself in the factory by being lower-cost labor.

And robot labor *is* cheaper! Ever since 1961 there have been ever greater economic advantages in robotics. Unfortunately, economic advantages alone have not really been enough to stimulate the kind of market growth that such benefits would seem to make inevitable. Somebody has to break the ice in a big way before industry will perceive robotics to be essential—when a competitor starts to make broad use of robots, the economic benefits will be obvious.

This was Unimation Inc.'s experience after having installed its first robot in a die-casting operation. The die-casters took a ho-hum view of this accomplishment. However, they began to take notice when one die-caster installed 10 robots, and they really took notice when this same manufacturer expanded to 35 robots unloading die-cast machines in one plant alone. This proved to be the breakthrough, and shortly thereafter Unimation was able to sell some 400 additional robots for use in the die-casting industry.

Today the greatest driving impetus comes from the competitive use of robots in Japan. The Japanese have made U.S. and European industrialists sit up and take notice of robotics.

"MUM's the word." Or at least that's what the Japanese say. MUM is their acronym for Methodology for Unmanned Manufacturing, eliminating human workers from the rote pick-and-place tasks. The goal, as established by MITI (Ministry of International Trade and Industry), is that humans will be "only knowledge workers by the year 2000." This is a dramatic objective—and one that Japan is unlikely to attain by the year 2000—but it does represent a worthy challenge for the balance of this century.

One way to reach the goal of MUM might be to simply replace all of the human drones with robots. In this scenario the workplace remains the same and the human roles remain the same, but everyone who is not a "knowledge" worker is replaced by a robot sufficient unto the dull, routine tasks at hand. The manufacturer would be "doing business as usual" except that he would be blessed with a cheap and docile blue-collar (steel-collar?) work force.

Many Americans visiting Japan are baffled by automation, specifically by robot installations for which they can't see any economic justification, that is, where there's no immediate prospect of payback. The Japanese quite often move to new technology that increases productivity strictly on the basis of a gut feeling that, in the long run, it must be right. The president of one large corporation noted that Japanese executives do not think in terms of the hourly rates for labor, but rather in terms of the lifetime cost of labor. A worker never hired represents a lifetime saving.

To reach prospective U.S. and European buyers of robots, however, the robot manufacturer doesn't have to rely on commitments based on intuition. Robot labor is already so attractive that it can be justified using any of the conventional economic formulas used to win over financial vice presidents. Once the "bean counters" are convinced that robots truly are general-purpose equipment and are not subject to short-term obsolescence, they relax and cooperate. The astute manufacturing manager gets his robotic automation and pockets the additional intangible benefits of throughput and improved quality.

We have seen that the deaf, dumb, blind, immobile, awkward, and slightly spastic robots of the '60s, '70s, and early '80s have found much work in industry. There is much more work for robots to do—and this would be true even if there were to be no more technological advances. Yet technology *is* dramatically on the advance. No self-respecting technical university is without its graduate robotics program, its pet project for enhancing robot attributes, and its programs to create roboticist Ph.D.'s.

For new jobs, all machines need is a new tool attached to their arms.
This robot has a circular saw as its working "hand."

One of the first tasks robot foremen assigned to the factory machine
was the dull, enervating work of materials-handling.

Furthermore, giant corporations in Europe, the United States, and Japan have established hefty research budgets to build proprietary robotic technology. Finally, the governments of most of the other industrialized nations have also joined the fray as robotics research sponsors and even risk-takers in the applications area. Already there are a number of applications coming into their own because of the output of this research and development. Robots with rudimentary vision, some tactile-sensing, mobility, and more powerful software are already proving themselves.

Batch Assembly These days the glamour industrial activity is assembly. Once again the Japanese are in the lead with robots being used in electronic and electromechanical assembly. A notable example is the assembly of videotape drives. Matsushita and Hitachi, for example, have very impressive assembly lines that crank out videotape drives at the rate of one every eight seconds.

While these assembly lines are impressive, the use of robots is really not very sophisticated. In the high volume lines, the robots usually have only two or three articulations and nothing in the way of sensory perception. We would like to see robots used more like human assembly operators. Batch assembly of a wide range of products would require full manipulative power of six articulations, vision, and touch-sensing in the robot hands. These capabilities are on the near horizon, and it seems likely that extensive use of robotics in batch assembly will be with us before the end of the decade.

As the various automatic technologies develop, it becomes more and more difficult to pinpoint what part robotics will play in the factory. As automation becomes more flexible, with all elements under computer control and responsive to a variety of external stimuli, robots may lose their distinct character. They may simply be distributed throughout the factory as moving elements of the entire production organism. Science fiction has given us a counterpart of this phenomenon. In *2001: A Space Odyssey*, HAL is a distributed robot that we never see in an embodiment but that permeates the space vehicle and whose evil intent cannot be aborted without the humans actually

With all the deftness of an octopus tentacle, this Swedish-built robot can reach into areas inaccessible to other machines.

crawling through the ship's entrails, chopping away to incapacitate HAL.

Sure, it's true that robot technology may one day endow a robot with the ability to pick up scattered parts in a bin. It's inevitable that robots will become more and more adaptable, but the factory of the future won't need this adaptability if all of the activities leading up to and on the factory floor are planned out. Even today's rather stupid robots can cope very well in an environment where everything has a place and everything is in its place. But, the "factory of the future" is not the only stomping ground for a robot worker. It is merely the first and the most obvious.

Suppose robots are given better understanding of the world in which they function; suppose they are made both mobile and fully aware of their workspace; and suppose they pick up all of the AI (artificial intelligence) capability that has become available, including the ability to communicate one-to-one with humans in a high-level language (such as English). What then are the prospects for robots?

No one should conclude that this is a blue-sky speculation. The applications we will now consider are already being physically explored. This is here and now, if we may be granted that "now" is all of this decade.

Someone who is familiar with the state of the art in the industrial robot industry, and therefore with all of the jobs that have been described in this chapter, might still look at the following list and snort, "Sheer science fiction!" But the fact is that serious experimentation is going on somewhere on every one of these jobs. It's important to note also that the ultimate worker in these jobs has left the factory and branched out into a spectrum of service activities.

Fast Food Preparation and Delivery The first effort at an automated restaurant was made over fifteen years ago by the AMF Corporation in a venture idea called Amfair. Evidently before its time, Amfair was unsuccessful. Today, however, innovators, armed with some remarkable technological advances, are almost sure to succeed.

In each of the Pizza Time restaurants, a chain of pizza parlors

with automated entertainers, robots are only motorized man-
nequins that entertain customers with taped routines, but maybe
this is a harbinger of the future. A working robot could be back-
stage making hamburgers, french fries, and cherry pies; a fellow
robot would then deliver the goodies to the waiting customers.
Wouldn't it be nice in that circumstance to have real, live, talented young folks to do the entertaining?

Animal Husbandry The Australians have taken a lead in the fascinating possibility of shearing sheep by robots. Both universities and private development organizations are attacking this
"harvesting" problem, which is the sheep rancher's one barrier
to independence. He is at the mercy of itinerant teams of sheepshearers, who move from ranch to ranch to accomplish this
arduous task.

Australia is, of course, a natural focus of this robotics opportunity, since the Australian sheep population is 140 million strong.
The research and development activity, of course, involves more
than the robot shearers themselves. The sheep must be marshaled and single-filed into a shearing station, and the robot
must be given some preliminary information about the size of
the animal being shorn. This is done by probing the beast in a
few critical spots and then using a generic program to create a
shearing pattern. The process is analogous to taking a standard
dress design and scaling it into patterns for various-size women.
But that is not enough, since some variability always exists.
Therefore, a robot arm for sheep-shearing must have the tactile
capability to guide the shearing head along the sheep's body
and yet not inflict wounds. Some cooperation on the part of the
animals is also necessary—they have to be restrained in some
way in the shearing station. One of the research groups paralyzes
the sheep by a low electrical current between nose and tail.

Nuclear Maintenance It's evident that a radiation-hardened mobile robot with on-board smarts and useful appendages could
be a great boon for tasks such as decontamination and repair of
hot places like the Three Mile Island power plant. Less dramatic
but equally vital is the use of robots in routine maintenance of
all of the nuclear power plants already in existence. Over the

A tabletop robot arm, the Unimate 900, does the delicate job of manipulating small and fragile electronic components with ease.

years some frailties have cropped up and human troubleshooters have had to go into areas where the radiation is hazardous to human health. Current procedure limits such exposures to five minutes, which then disqualifies the worker for another such assignment for up to two months. Naturally, this is an extremely expensive procedure.

ROSA, Remotely Operated Service Arm, is the Westinghouse Electric Corporation's entrant in the nuclear maintenance sweepstakes (sweepstakes because other manufacturers are hot on the trail of Westinghouse in attempting to develop robotic nuclear servicemen, among them a consortium of Japanese companies). ROSA is currently at the forefront—a number of these radiation-hardened, six-articulation mobile machines exist and already are on the job. They are armed with an arsenal of specialized tools, and they can work either independently under program control or as manipulators under the control of remotely located humans monitoring ROSA's work with television cameras.

Hospital Aide Here again the Japanese are apparently being most aggressive in developing robot technology. The Japanese particularly point out the advantages of a robot aide to lift an immobilized patient while bed linen is being changed. That same robot could aid the patient in bathroom visits. The potential tasks are legion. Robots, particularly mobile robots, could deliver food and medicine on schedule and perhaps be busy between meals scrubbing hallways. More sophisticated robots could extend domestic chores into cleaning and disinfecting rooms, wards, and toilets.

Neurosurgery Whenever I mention neurosurgery as a potential robot application, the idea is usually greeted with incredulity. Folks say, "Now you have finally gone too far." But the fact is that sometime within the next few years a robot will be part of a brain surgery team at Memorial Hospital Medical Center in Long Beach, California. The robot's job will be to help the neurosurgeon in treating what is considered to be an inoperable brain tumor. The procedure will first be to find the tumor with CAT scans interpreted by a computer to establish the precise location inside the skull. A robot fixed in relation to the patient's immobilized head is then told by the computer where the tumor is and the location that is best for drilling a hole in the cranium. The robot moves to this location, locks its joints, and then informs the surgeon that it is ready. The surgeon directs the robot to drill the hole, and he may then insert a tube to a specialized depth to drip radioactive medicine onto the tumor. This procedure now requires five to six hours of a neurosurgeon's time. With the unique assistance of a robot paramedic, a neurosurgeon will be able to carry out the procedure in thirty minutes.

Household Servant What better ultimate worker than a personal slave? The simple robot jobs now being done by robots, the more sophisticated jobs slowly being assumed in factories, and all of the service jobs that have just been discussed are creating a rich and versatile technology. Before the end of the 1980s, that technology could be marshaled to create a household servant. The house will have to be redesigned to welcome the robot just as automated factories have, but it will remain a comfortable home for its human inhabitants even if they must defer to the needs of their robot servant.

There will be a robot pantry. In it will be the robot's central intelligence and its communication links with the balance of the household as well as a data link connection with the outside world. The robot itself would be mobile so that it could carry out its assigned tasks such as cleaning the house; washing the windows; setting the table; cleaning the table; washing dishes; serving guests; entertaining children; protecting the house from fire, flood, and intrusion; and maintaining the numerous automated appliances, such as stoves, dishwashers, centralized vacuum cleaners, furnaces, and air conditioners.

As one who has spent a large part of his working career in developing robots, I find that the creation of an ultimate worker for my personal service is indeed a pleasing prospect for what I hope will be a comfortable, stimulating, and even companionable retirement.

Of course, there is still time for Joe Engelberger to participate in the creation of this robot, but just in case I cannot do it all before my retirement, I am saving up to be one of the earliest customers. I know that the household servant will become a reality, and I am also realistic enough to know that it will be a big ticket item when it first enters our lives.

The story of the ultimate worker wouldn't be complete if we simply adhered to the role of the robot stepping in for human workers in normal working environments. We may also expect robots to go where all men fear to tread. They may live in poisonous atmospheres and at temperatures intolerable to man. They may go into the depths of the Earth, not for a shift but forever. They may be the only creatures able to cope with the miscalculations of mankind, such as Three Mile Island. It is conceivable that they will be more than the ultimate workers, they may become the ultimate survivors should humans become so intemperate as to destroy our Earth's life-support systems.

But there will surely be social scientists saying, "Never mind the far-out, evolutionary potential for robotics. Let's instead look at the real, short-range, negative effect of robots." Their arguments always come down to fear of crippling human society by taking over all the gainful employment.

In any debate with a social scientist, my main position is: "Any gain in productivity is always good." I take this position with almost religious fervor. I call upon history as my witness, and I challenge any social scientist to imagine a scenario in which this axiom does not apply. By productivity I mean efficient use of natural resources, capital, and human labor. Automation can help us to use our natural resources and capital more effectively. However, most people tend to equate automation not with the more effective use of human labor but the outright displacement of human labor. It is this point that has many social scientists crying, "Foul!" No one can deny that dislocation follows as a result of technological advance. Such has always been the case, but in an imperfect world social scientists would be well advised to test an emotional conviction against a more rational cost/benefit analysis. Over the long run, do benefits outweigh the costs?

An example: In 1870, 47 percent of the United States' population was involved in agriculture. By 1970, this percentage had dropped to 4 percent, yet the agricultural industry was creating vast surpluses. Did this tremendous gain in productivity cost something? Of course, but how could anyone make a case that the costs exceeded the benefits we enjoyed? The same argument applies to our creation of material wealth through the automation of manufacturing. There will be dislocations and there will be pockets of distress and of unrest, but the benefits to society will overwhelmingly exceed the costs.

Why is it always assumed that increasing productivity through technology must result in increased unemployment? Who says that the forty-hour week is sacred? Why not a twenty-hour week, if this is sufficient to provide us with material comfort and security? Suppose we increase our productivity at such a pace that we would have sufficient excess wherewithal to clean up our polluted waterways and atmosphere? To rebuild our bridges and highways? There are no ends to the directions in which we could usefully apply human energy were we productive enough to free some of this energy and wise enough to distribute our material output usefully. Distribution of wealth is the question the social scientists should concentrate on—and more importantly, our politicians, some of whom will, we may hope, develop into statesmen. Let the roboticists press on, goad them to greater success so that the wherewithal is there for distribution.

In the year 1984, George Orwell's *1984* once again became a bestseller. He was a remarkable forecaster. In 1948, he predicted data banks with detailed personal information on everyone. He predicted think tanks where experts would plan future wars. He predicted poisons capable of destroying vegetation à la Agent Orange. He conjectured about genetic manipulation in the creation of disease germs that were immunized against antibodies. He foresaw a lack of heating fuel and electricity, and he even predicted a distressing merging of the genders.

In his commentary on *1984*, Erich Fromm, philosopher and psychiatrist, said: "George Orwell's *1984* is the expression of a mood and it is a warning. The mood it expresses is that of near despair about the future of man and the warning is that unless the course of history changes, men all over the world will lose their human qualities, become soulless automatons, and will not even be aware of it."

There are many who look at our circumstances in the year 1984 and, like Fromm, "despair about the future of man." Still, it's just possible that our destiny is not to become "soulless automatons." George Orwell was very well read and had to have known about the robot concept. *1984* was preceded by Czech playwright Karel Čapek's *R.U.R. (Rossum's Universal Robots)* and some of the early stories in Asimov's *I, Robot* series. Yet among all of his dire predictions he never found a place for robots. I submit that this was crucial to his thesis. If humans were to become automatons, there would be no need for a mechanical replica. Human automatons would be more cost-effective by far for getting the subhuman work done.

But today, in the real world, we do have robots. Their population is growing ever faster and their capabilities are becoming ever more impressive. From this we can draw hope that Orwell will ultimately be proven wrong. *Robots* will be our automatons. They will take over the dirty, grubbing subhuman tasks, leaving us humans and our souls gloriously intact.

The Machine Servant
Richard Wolkomir

So inexorable is the evolution of industrial robots that household spinoffs—mechanical slaves—are virtually inevitable. You can already install a simpleminded robot in your home to serve highballs to your guests.

The 1983 graduation speaker at Anne Arundel Community College in Maryland only 4 feet tall and pudgy, was not prepossessing. He arrived on stage with an awkward lurch, looking faintly ludicrous wearing a formal black bow tie. He had no advanced degrees, no resume of stunning achievements. Nevertheless, Robot Redford mesmerized the graduates, for he was history's first commencement speaker made of fiberglass.

It was not that Robot Redford spoke scintillatingly—he offered a few not-so-pithy remarks on the need for humans to work with robots and technology to solve society's problems. But, like the proverbial talking dog, what he said was less noteworthy than that he was speaking his piece at all. In fact, Robot Redford's 175-pound white body, a TV camera snout atop his chrome head, a video screen embedded in his chest, symbolized a revolution—already begun—that will transform the way we live: the resurrection of slavery.

Abraham Lincoln, who said, "He who would be no slave must consent to have no slave," need not roll in his tomb. The new slaves, just beginning to edge into our homes, will be the descendants of industrial robots, things of aluminum and transistors. We are about to discover the exquisite pleasure of equipping our homes with ambulant entities whose only reason for being is to do what we tell them to do. We can be polite with them, cross, masterful, or demanding, and we need never feel the least tweak of conscience, for they will be mere mechanisms, perfectly obedient, but unfeeling, soulless.

Spooky? At first, we may have difficulty accepting these electronic strangers. For instance, Robot Redford is only an extremely primitive version of the robot servants to come, merely a novelty and a toy. Nevertheless, before graduation, the Anne Arundel Community College campus was in an uproar over the propriety of having a fiberglass graduation speaker. As Kathy Hammac, the class valedictorian, later told reporters, "At first I thought it was inhumane and somewhat degrading to students. But the robot didn't take away from the ceremony. It represents a look into the future, something our generation is going to have to deal with and handle in the years to come." "It was delightful! It was beautiful!" said a graduating music major, age sixty-three. Another graduate said, "You forget a political leader, but you won't forget a robot."

Robot graduation speakers are only a hint of what is coming. So inexorable is the evolution of industrial robots, say the experts, that household spinoffs—mechanical slaves—are virtually inevitable. In fact, you already can install a simpleminded robot in your home, perhaps to serve highballs to your guests.

An Urbana, Ohio, computer engineer named Charles Balmer, for example, has created Avatar, one of scores of robots now rolling out of the basement workshops of electronically savvy hobbyists. Resembling a cross between R2D2 and a dental chair, Avatar contains a 64K computer and electronic sensors that enable it to navigate through a home. However, if you do not feel up to designing and building your own robot, you can buy one ready-made.

One over-the-counter model, Topo, a "personal robot" that has been manufactured by Androbot, Inc., of San Jose, California, costs approximately $500. It is 3 feet high and rotund, earning the nickname "plastic snowman." Topo, customers agree, is cute. But it has a major drawback: It doesn't do much. Basically, Topo can move forward, backward, right, and left. Most of its sales have been to schools and colleges, as a teaching tool. Oregon State University, for instance, uses one of the 50-pound automatons to teach schoolteachers programming in both LOGO and BASIC. However, Androbot's guiding spirit, Nolan Bushnell (the flamboyant entrepreneur who earlier founded Atari and popularized video games), announced that Androbot will be marketing new, more sophisticated designs. One, according to Bushnell, will be B.O.B. (Brains On Board), a much brighter version than Topo with its own on-board computer. Retailing for about $2,000, B.O.B. will accept software packages to fit the tasks you want it to perform, scaling its price up to $4,000. Said Tom Frisina, then Androbot's president, "It will be cartridge-driven and real user-friendly, so that even Joe Six-Pack would want to buy it." With its twinkling red eyes (actually infrared sensors), B.O.B. will be able to move around the house, teaching languages, playing games. At a signal from your infrared "beacon," B.O.B. will roll off to the kitchen, shooting out infrared and ultrasonic signals to avoid hitting the walls and the sneaker your five-year-old left lying on the floor. B.O.B. will roll up to a specially designed refrigerator. The refrigerator will pop out a beer and B.O.B. will trundle it back to where you sit in the living room chair, watching TV.

Other robots are already on the market, eager to serve you, in their limited way. For instance, the Heath Company, of Benton Harbor, Michigan, makes Hero I, a 39-pound, 2-foot-tall midget that can utter thirty-nine different phrases, including "I do not do windows." On the other hand, Hero *can* do lobbying in Washington. In 1983, Hero appeared before a Senate subcommittee and told the bemused legislators, in its electronic voice, that they ought to finance industrial research. Just to show off a bit, Hero sang a verse of "Happy Birthday." Then, with its one metal arm, it handed out copies of its testimony.

The robot costs $1,500 as an assemble-it-yourself kit, $2,500 if you buy it prebuilt. For that price, you buy a shrimpy robot that can carry up to 16 ounces in its clawlike gripper, check rooms for unexpected movement, deliver drinks, wake you at a given time, and handle similar tasks for which you program the on-board computer. Hero can perform other duties. For example, it can detect a burglar breaking in, raise itself to its full 2 feet, extend a threatening gripper, and announce in tones sure to straighten the hair of a hardened crook: "Warning—Intruder! I have summoned the police!"

And there are other, smaller robots. Mitsubishi Electric sells

By the middle of the next century, the robot fetching the morning paper might be a common household event. Already a few companies, such as RB Robot Corporation, are making what they now call "personal robots," like this RB5X, commercially available.

a desktop model, with a claw that shuffles your papers and answers your telephone. Spectron Instruments, of Denver, sells a programmable arm, with rudimentary vision, in kit form.

More ambitious is RB Robot Corporation's 2-foot-tall RB5X, whose on-board computer and sensors enable it to serve you a drink or hand you your *Wall Street Journal*. Made in Golden, Colorado, what looks like an animated crockpot can sense intruders and play "spin-the-robot" with the kids.

What none of the domestic robots now on sale *can* do, however, is anything really useful. So far, they are little more than sophisticated toys. Yet they are forerunners of the truly useful robots to come, a direction in which the housewares industry already is groping. For instance, Hubotics, Inc., of Carlsbad, California, announced in 1984 that it would soon begin selling what it calls "the ultimate appliance," named Hubot. Here are some of the things Hubot can do: wake you in the morning by tapping your shoulder; bring your morning cup of coffee and your newspaper; play its on-board radio, cassette player, or TV; tap into data banks; allow you to write notes on its built-in computer workstation (complete with printer); turn the lights on and off; serve as a burglar *and* fire alarm; play video games with your kids; turn on the dishwasher; vacuum the carpet. At least, so promises Hubotics, Inc., and all for around $3,500.

Robots seem to be evolving on two fronts. One is in the design of hobby robots, teaching robots, and the early personal robots of only dubious usefulness, like Hero I, Topo, and RB5X. The other front is the growing sophistication of working robots, now beginning to spread from factories, where they perform repetitive chores like riveting and painting, to more demanding non-factory jobs. Eventually these working robots will grow sufficiently adroit for household and other use.

For instance, in a number of towns today, such as Stamford, Connecticut, a system called the Telso Programmable Robot Telephone Operators is cutting down on truancy. When students are late or missing at Stamford's 1,900-student West Hill High School, the $5,000 robot dials their parents, making calls every weeknight after dinner, with never a request for overtime pay. The machine delivers a 45-second message, telling the parents that their children have been truant. If there is no answer, or the answerer hangs up before the message is completed, the machine calls back twice. School officials say the robot caller has reduced absenteeism approximately 9 percent. Students call it the "Voice of Doom."

And then there is Denny, a 4-foot-tall, 400-pound robot resembling a high-tech garbage can, but actually a prison guard. Denning Mobile Robotics, Inc. of Woburn, Massachusetts, has contracted to produce up to 1,000 of these electronic guards for Southern Steel Co. of San Antonio, Texas, the nation's largest manufacturer of prison security systems. At $30,000 each, the robots will outperform human guards at some dangerous or boring jobs, according to Denning's marketing vice president: "The robot can maintain the same level of consciousness—it doesn't get sleepy or careless," he says.

Denny will roll along the prison corridors at 3 miles per hour on its three wheels, sensing its way with a sonar rangefinder developed by the Polaroid Corporation. At the same time, other sensors will shoot out infrared and ultrasonic beams, hunting for unauthorized intruders, all the while transmitting its findings to a human-run central control room. If the machine detects an intruder, it issues a stern warning: "You have been detected!" It also can ask such questions as, "Who are you?" Although weaponless, the roboguards will, as one company official puts it, be durable enough to handle "the hard knocks of prison life."

Of course, homes are neither prisons nor truancy-plagued schools. But the technologies enabling robots to dial phones and deliver messages and patrol prison corridors also will enable them to call the grocer or alert the local patrol to watch for burglars. Thus, the evolution of industrial robots will eventually lead to the birth of the household robot.

Technologies like Denny the Guard Robot are steps toward household slaves, and the home appliance industry already is taking note. For instance, Elliott Wilbur, an expert on housing and a vice president at Arthur D. Little, Inc., the international consulting company headquartered in Cambridge, Massachusetts, says that, although he cannot reveal the details, one of his firm's big-corporation clients is currently experimenting with robots designed specifically for household use.

During one meeting with eight other Arthur D. Little experts in fields ranging from electronics to home appliances, Wilbur and his colleagues analyzed household robots as a potential product. They disagreed, often sharply, on exactly how robots would fit into the home appliances market. But they agreed on one important point: Developing the necessary technology is not only feasible, but virtually inevitable. As engineers steadily boost industrial robots' IQs, they are creating the technological bits and pieces that eventually will come together into household "slaves" that will feed your Siamese, rake your leaves, and knot your cravat.

Not that it will be easy. Housebroken robots, if they are not to exit rooms through the wall or shatter the goblets in which they are serving burgundy, will require a number of senses and abilities that today's industrial robots lack. And developing those senses will be arduous. Consider just one—vision.

Donald L. Sullivan, an Arthur D. Little computer expert, is developing a robot inspector for industry. His machine looks at a product, such as a slice of bacon, through video eyes. It decides if the slice has too much fat, either passing the slice along or tossing it in the reject bin. The system operates by translating the meat's lights and darks into numbers a computer understands.

"Producing a digital image of merely a slice of bacon takes about 76,000 numbers," says Sullivan.

His laboratory is a creative jumble of video cameras, microprocessors, and—a Salvador Dali touch—slices of meat, pouches of sweet-and-sour pork, and crackers. Robot vision, he says, works by translating a video image into a computer's numerical language. The system assigns a number to each shade of gray between black and white—the lighter the shade, the higher the number. Then, breaking the image its video eyes see into dots, it assigns the appropriate number to each dot. For the image of a bacon slice, for instance, the computer is programmed to identify all dots with values below a certain number as dark background, above a certain number as white fat, and in between as red meat.

Compared to moving through a house, it is simple for a robot to inspect bacon for fat. Industrial robots, after all, repeat the same tasks in a controlled environment. Yet, just developing one experimental inspection robot, says Sullivan, has so far cost about $70,000 in hardware and $300,000 in engineering time.

Like Sullivan, researchers worldwide are hard at work im-

Would you want your guests greeted by this machine?
Some of the early models of robot help
may resemble this demonstration "showbot."

proving robot vision, for a sharp-eyed machine would have myriad uses. One project, a joint effort of the University of Florida and the Martin Marietta Corporation, is adapting missile guidance systems to the more peaceable task of picking oranges. "In very simplified terms, the same thing that helps a missile find a tank could help a robot arm find an orange," the university's Chairman of Agricultural Engineering, Gerald Isaacs, told reporters in announcing the new project.

First, the robot orange picker would view the tree through its optical system and store the image in its computer for analysis. The computer, having processed a "map" of the tree, would then guide the machine's robot arm—which would have its own independent video optical system—to pick the oranges that are ripe. A Martin Marietta official says the machines should be a common sight in the groves in the 1990s. He says the machines will work at night because the robot's circuits will operate better when it is cooler and because, under artificial lights, technicians can fine-tune the illumination to the needs of the robots' vision systems.

Industry experts say that, by the 1990s, 35 to 40 percent of all industrial robots will have "eyes." Object Recognition Systems, Inc., of New York, has already developed a prototype of a seeing robot that can study a bin full of pens, analyze the jumble, and then—deftly reaching in with its mechanical claw—pluck out a single pen.

Meanwhile, at the University of Massachusetts, researchers are working on a more advanced system that will see the natural, outdoor world—recognizing colors, textures, three-dimensional placement, even applying the correct names to the things it sees. In addition, the system will be able to compare what it sees to "schemas" for how the world operates, stored in its memory banks.

According to Dr. Edward M. Riseman, one of the University of Massachusetts computer scientists developing the new system, "We could have motion, touch, and static vision all working together to build a three-dimensional representation of an object, label it visually, grasp it, and manipulate it in manufacturing processes." Of course, the same technology applies to domestic robots, as well.

But robots also will need other senses, in addition to vision.

Touch, for instance. And that, too, is on its way, courtesy of industry. According to Leon D. Harmon, a professor at Case Western Reserve University, electrical engineers will be able to simulate a sense of touch in robots by placing pressure-sensitive pads at the ends of their fingers. The hard part, he says, is developing computer programs that let the robot know what it is touching. He is trying to give robots the ability to sort objects and recognize the feel of the correct part, as well as how it is oriented, before picking it up.

Scientists at Japan's Nippon Telegraph and Telephone Public Corporation recently demonstrated a robot that can turn the pages in a book. The robot can turn one page at a time or any number, depending on its instructions. Says one of the researchers, "It first gauges the size of the book and the thickness of the pages. It can then work quickly and accurately, and we guarantee there won't be a torn page in sight." Uses for the robots, which can read aloud from the books, as well as turn the pages, will range from radio newscasters to bedside readers for the blind and disabled.

Right now, robot touch is still rudimentary, ranging from the simple, like the little button that snaps on the refrigerator light when you open the door, to the more refined touch of today's factory robots, which have sensors in their claws to suggest how hard they are gripping an object. But the next generation of touch-sensitive robots will be able to "feel" the shapes of objects and identify them.

At Carnegie-Mellon University, researchers have developed a rubber sensor pad that can distinguish among six objects, ranging from a battery to a wrench. At the Massachusetts Institute of Technology's artificial intelligence laboratory, researcher John Purbrick is developing a 2-inch-square pad containing 256 electronic sensors. The sensors are so exquisitely sensitive that Purbrick can press his fingertip against one and watch the image of his fingerprint appear on a video screen wired to the sensor.

For researchers like Purbrick, the goal is an industrial robot sufficiently sensitive to install typewriter springs or know that a bolt it is holding is upside down. But our aluminum household slave will also need a light touch if it is to wash our crystal stemware, or the dog. Such technologies will develop in industry, then reach the home.

Seeing and a delicate sense of touch are vital, but our household robot slave also must hear and understand our commands—a formidable challenge, considering our myriad regional accents and idioms. Also, we each have personal speech idiosyncrasies: lisps, stammers, pronunciational peculiarities, and odd habits, such as trailing off sentences uncompleted. The robot also must factor out such static as sneezes, coughs, snapping gum, and the teenager next door squealing his Camaro's tires. Nevertheless, the technology is virtually inevitable because the payoff will be so huge in the office automation market.

The ponderous workings of speech recognition systems, if nothing else, underscore how deft is even the dullest of human brains. First, the machine—via a microphone—converts the speaker's voice into an electrical signal. Then it converts the signal into a digital code for analysis by a computer. The computer then translates the code into a computer-comprehensible message, in which each word becomes a string of numbers, representing the word's sound spectrum. Stored in the computer's memory are the strings of numbers representing each word in its vocabulary. When it "hears" the speaker, it compares his digitized words to the strings of numbers in its memory, thus recognizing speech.

IBM researchers already have developed a typewriter that takes dictation, typing out what it hears. Still experimental, the system must be "familiar" with the voice of the operator, who speaks into a microphone. Using an acoustic processor program, the machine breaks down the words it hears into sound patterns. Then another computer program, called a linguistic decoder, analyzes the sound patterns and translates them into a sentence, which appears on a computer screen.

So far, the system has attained about 91 percent accuracy. However, while this system may be "smart," it is far from brilliant. It has a vocabulary of only 1,000 words and can understand only one person at a time—and needs 2 hours to memorize each speaker's speech patterns. Worse, thinking over what it has heard takes the machine a long, long time: To analyze and print what takes you 30 seconds to say, the machine requires an hour and 40 minutes. In other words, if it were mounted in your personal robot right now, and you said, "I smell something burning— turn off the stove," the house might be a cinder by the time

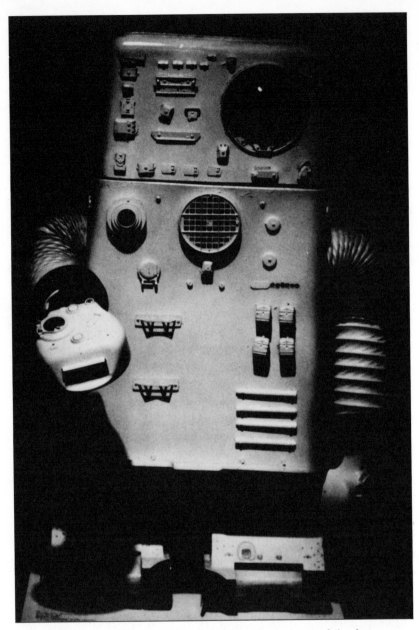

Another speculation on what form the robot servant of the future might assume. This model is "Bumpy," a hand-built personal robot.

the robot finished mulling over your command and trundled off toward the kitchen and the stove.

Nevertheless, such research will ultimately lead to machines with ears to hear and minds to comprehend. There are already some factories where quality-control inspectors report their findings verbally into speech-recognition machines. Developing a robot smart enough to do what it's told, and do it fast, will not be easy. Nevertheless, as David Lee, an Arthur D. Little expert on consumer products and appliances, puts it: "Voice recognition is well on its way."

Meanwhile, a household robot should be able not only to hear and obey, but also to speak: "Dinner is served, Master." However, robots already can speak better than they can listen. Right now, for instance, the "operator" giving users of New York City pay phones such messages as "Sixty cents, please," is a computer with a seventy-word vocabulary, including "Thank you." You can now buy a talking clock that literally tells you the time, a talking camera that tells you when to use your flash, and a car that tells you, "Please turn off the lights." It will not be long before machines are chattering away all around us.

Engineers have developed several techniques for giving tongues to machines. They range from actually modeling the human vocal apparatus to storing the basic elements of speech in a computer's memory and putting the elements together into words or sentences according to the rules of language construction. Already over a dozen computer chips for generating speech are either on the market or in the design stage. A few decades hence, when you tell your household robot to change the bulb in the bedroom lamp, there is every reason to expect it will respond with, "At once, Your Magnificence—60 or 100 watt?" But don't expect it to run off to perform its humble chore; it is much more apt to roll.

A walking robot is far from impossible. Robert B. McGhee, an Ohio State University electrical engineer, has built a six-legged walking machine with sensors in each foot. This metal cucumber beetle can even pick its way along a path littered with stumps. At Carnegie-Mellon University, professors have even choreographed a dance for a robot and a woman. Nevertheless, the first domestic robots are not likely to lurch through your house on metal legs. "The wheel was one hell of a good invention," says one expert.

One last sense that your robot may have, incidentally, is a sense of smell. Carnegie-Mellon scientists are also working on technologies that will give robots a nose more sensitive than a human's. "We feel the need for robots to be able to sense all the kinds of things we can sense and then some," says Mel Siegel, a researcher at the university's Robotics Institute. "I'm certain it can be done," he adds. "The problem is the sense of smell has a lot of chemistry in it, a lot of physics in it, a lot of biology in it, and a lot of psychology in it as well."

To create a computerized nose, Siegel and his associates are studying the simplest smells to start, such as ammonia, methane, and ethyl alcohol. He says that such substances have far simpler aromas than a substance like gasoline, which has 300 to 600 components in its scent, most of them detectable by the human nose. A lemon peel's aroma has 400 separate elements. Siegel hopes to combine devices sensitive to various chemicals into a single electronic sniffer, a multilayered computer chip. For each odor it encounters, the chip will generate a unique electronic signal, which a computer will identify.

Carnegie-Mellon's work on the robot nose is sponsored by industry, which could use such systems for detecting drugs at airports, helping police detectives uncover clues at crime scenes, preparing food in restaurants, and warning miners of poisonous fumes. However, such a system also would help a household automaton as it goes about its chores. For instance, if it sniffs smoke, it will know it should turn down the burner under the eggs it is frying for your breakfast.

Industry, then, is already underwriting research on the senses and abilities a domestic robot will need: vision, speech, speech recognition, touch, locomotion, even smell. And the experts seem to agree on when all these bits and pieces of new technology will come together in the first robots truly useful in the home. Arthur D. Little, Inc., consultants, in their roundtable discussion on the advent of robots into the home appliance market, predicted that commercial models for the average buyer will probably debut a bit later, about the year 2000.

But the $64-billion question is this: Will anyone buy the things? As Arthur D. Little expert Elliott Wilbur put it at the Cambridge meeting: "Why not just hire a kid to cut your grass?"

The issue may be moot. Electronics engineer Stuart Lipoff, another participant in the roundtable discussion on robots, noted: "We already have programmable microwave ovens, dishwashers, and swimming-pool cleaners, which are all robots of a sort, even if they don't have eyeballs."

The consensus at the meeting was that robots will develop along two tracks. First, our appliances will get smarter and smarter. "The cost of braininess is low," says Martin Ernst, a vice president of the consulting company and an expert on operations research. Eventually, we'll begin to merge the smart appliances. "You might end up with a unit that combines a refrigerator and an oven," said appliances engineer David Lee. At a preselected time, the freezer section would pop the dinner you punched in for tonight into the microwave section, which will turn itself on. The next step, according to Lee, will be an automated menu—choosing your dinners for the next month, perhaps—coupled with automatic inventory control based on supermarket package codes.

Meanwhile, on robot evolution's second front, engineers will be developing an independent, stand-alone automaton to handle specific odious chores, like leaf raking and cleaning the bathtub. Ultimately, the stand-alone robot and the smart appliances might share a brain, with one central processor controlling everything. "You have the mechanical peripherals, with modest intelligence, and a basic computing engine that ties everything together," says Gordon Richardson, an Arthur D. Little robotics consultant. In other words, some day, your house itself may be a robot.

In fact, the robot slave may take a more bizarre turn than at first seems likely. After all, we already have cars that talk to you, reporting that you need more gasoline or that you left the key in the ignition. It is not impossible to imagine coming home after a long day at work and telling your door, "It's me," whereupon it unlocks. And, inside, you might tell your house, "Rough day—get me a scotch on the rocks, and what's for dinner?" Your house, of course, would respond.

Many of the technologies for such a robot house already are here. NASA's Space Shuttle, for instance, is largely computer-controlled. Several Japanese factories already operate almost totally with robot labor under the control of computers, with

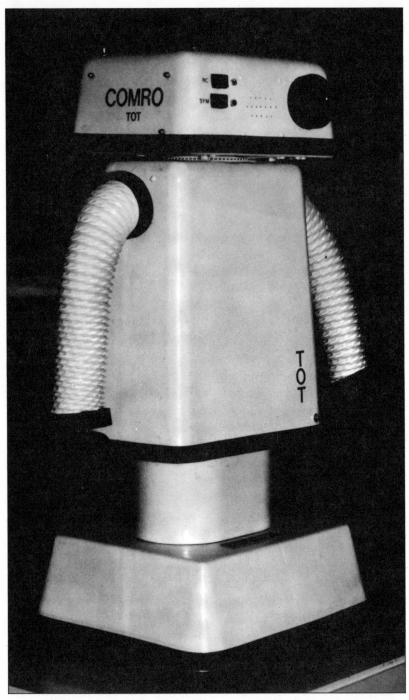

TOT, its makers say, will do carpets, act as a sentry, tell jokes,
or fetch things either by radio control or by its own computer programs.

only one or two humans on hand to monitor. Under construction in a Tokyo shipyard is a 170,000-ton freighter with a computerized system allowing one man to run the entire ship by voice commands.

Builders and inventors are already beginning to apply such technologies to the home. In Greenwich, Connecticut, the Copper Development Association has built the computer-controlled Sun/Tronic House, in which computers operate the shutters for the passive solar heating system and switch from the electric company's lines to the house's own batteries, charged by photovoltaic cells. The computers also do everything from telling you if you have enough steaks left in the freezer for Saturday's barbecue to reminding you of appointments.

The first customers for tomorrow's robot houses may be the elderly and invalids. Addressing the British Association for the Advancement of Science not long ago, bioengineer Heinz Wolff, head of the Clinical Research Center at Northwick Park Hospital, said there is no reason we cannot build housing for the elderly that would act as robot helper. Such a talking house could, for instance, remind its inhabitants to remove a kettle from the stove when the water inside is boiling. "It could even remind them to take their medicine," he said. "Just to illustrate how seriously we are taking this, we've gotten to the point of discussing what kind of voices one should use on this." The house's voice could be the inhabitant's own voice, for instance. Or it could be the voice of a favorite grandchild. The voice reminding inhabitants to take their pills might be their doctor's voice.

General Electric is now developing a home automation system. It is not quite a robot home, but it is a first step in that direction. At its Portsmouth, Virginia, laboratories, General Electric is developing a computerized system that will control a house—all appliances, burglar alarms, and heating and cooling systems. Via your TV and your house wiring, the computer would send directive impulses to appliances and monitor their status, giving you such vocal messages as, "The laundry is ready." If you are away, you can call your home on the telephone and change the computer's instructions for the heater, air conditioner, microwave oven, and the lights. In an emergency, the system would telephone the police and the fire department; then, if you're away, it will call you.

According to Joseph Engelberger, whose company has built a prototype of a domestic machine, the robot house will have a "servants'quarters." There, the household automaton stores spare parts for all the appliances in the house, along with a collection of tapes giving maintenance instructions. "At night," he says, "you'll tell the robot, 'The range isn't working, so please fix it by morning.' " While you sleep, the robot is "awake." It is alert for intruders and fires, of course, but it also is working on your stove. "If it gets stuck, it calls the factory and talks to a smarter robot and finds out what to do," says Engelberger. If it lacks a part, it orders it. By the time robots have advanced to this stage, they also will keep your larder stocked, ordering replacement items from the supermarket as needed.

Whatever form the robot takes, safety must be engineered in. "You'd hate to have a 2,000-pound robot go berserk in your living room," says robotics expert Gordon Richardson. In 1981, an autoworker at Japan's Kawasaki Heavy Industries was killed by a robot. The machine was malfunctioning; the worker, Kenji Urada, leaped over a chain fence to check it, although company rules called for opening the fence's gate, which would have deactivated the robot. Still moving, the malfunctioning arm swung, crushing Urada against another machine while his co-workers looked on helplessly.

In 1984, Litton Industries agreed to pay $7 million to the family of a Detroit autoworker who was killed by a Litton-manufactured robot at a Ford Motor Company factory. Because the five-story robot that normally did the job was malfunctioning, a supervisor had told the worker to climb to a rack to retrieve some castings. When the man was on the high rack, however, the robot moved, striking his head and killing him. Thus, before robots are ready for home use, certain safety precautions will have to be programmed in.

Household robots will have to be at least as safe as highly trained guard dogs, and domestic robots will certainly include in their programming the Three Laws of Robotics, propounded decades ago by Isaac Asimov. But will tomorrow's household slave necessarily be a machine? "The ultimate answer," says Elliott Wilbur, "may be to implant this intelligence in an animal, like a monkey." He suggests a microprocessor collar pulsing out signals. Nor is the idea of zoological slaves farfetched, considering that for eons mankind has exploited the muscles and brains of beasts from llamas to sheepdogs. At the Arthur D. Little meeting on household robots, computer expert John Langley cited his milkman, who recently became one of the last Boston-area deliverers to switch from a horse-drawn wagon to a truck. (Ironically, the milkman said that he found his old horse to have been more efficient than the truck, since the horse directed itself down the street and knew all the stops. "The milkman just ran along behind the cart, carrying bottles to his customers' doors," said Langley.)

Researchers at the Tufts-New England Medical Center Hospital in Boston, working under a National Science Foundation grant, are currently training monkeys to perform such services for paralyzed people as fetching food from the refrigerator, opening or locking a door with a key, removing a record from its album cover and placing it on the turntable, and brushing their owners' hair.

Martin Ernst points out that, with the population aging, such services will increasingly be in demand. And another Arthur D. Little engineer, Richard Whelan, notes that nurse robots could help elderly patients remember to take their pills, monitor life signs, and alert medical services in emergencies.

"Machines like these would allow people to remain independent and function on their own much longer," says Whelan.

It may well be that the handicapped will become the first owners of household robots. However, robots for general home use are sure to quickly follow the nurse robots onto the market. For both types of automaton, one question looms: How readily will humans accept robot servants?

Nobody knows for sure, of course, but there are indicators. People often anthropomorphize robots. In Japan, where the robot revolution is further advanced than in any other country, most humans accept the machines enthusiastically. A recent *Congressional Quarterly* report quoted Tokyo psychologist Seiichiro Akiyama on the phenomenon: "We give them names," he said. "We want to stroke them. We respond to them, not as machines, but as close-to-human beings."

So positively do many humans respond to robots, in fact, that the first domestic automatons may be less servants than friends. Psychologists and sociologists have been tracking what appears to be a growing epidemic of loneliness in Western society, caused by such factors as the high divorce rate, women tending to out- live their husbands, and the tendency of many jobholders to move to communities far away from the towns where they grew up, and where their families and friends still live. Thus, many of us may buy our first robots mainly to serve as chums.

In his book *The Intimate Machine*, Welsh psychologist Neil Frude says there is no reason that a computer—a robot's brain— could not be programmed to serve as an ideal companion. It would have a comfortable, informal style of conversation, but a certain unpredictability would be built in, to keep the machine interesting. When you first bring it home, it might be a bit hes- itant and stiff, just like a new human friend. Eventually, how- ever, it would warm up, exchanging ideas and information with you. It might have a sense of humor, too. "An artificial rela- tionship of this type would provide many of the benefits that people obtain from interpersonal friendships," says Frude.

"I think the first commercially viable item may turn out to be a robot pet, eventually even a robot lover—don't forget the 'orgasmatron' in Woody Allen's movie *Sleeper*," said electronics engineer Stuart Lipoff at the Arthur D. Little roundtable dis- cussion on domestic robots. He said the first models would be fairly simple, with some artificial vision, some artificial voice, the ability to understand speech, some movement. "It wouldn't have to do much more than move around, blink its lights, and respond in a playful way, maybe wag a tail," he added. Such robots might be therapeutic in nursing homes where patients are denied pets. Soft and fuzzy, they could have built-in heating units, making them warm to the touch.

Will people really choose machines to be their buddies? "Go back to *Star Wars*," suggests Lipoff. "What were those two robot creatures really doing? They were not so much utilitarian as companions, friends."

You could do many things with your robot friend. Certainly you could play chess or checkers. And, as with any friend, you could have long personal talks. The machine would never tire of hearing your problems and thoughts.

"If people already have trouble differentiating between their relationships with an actual psychologist and a computer ther- apist, over the next decade or so, as we develop computers with ultra-high-speed parallel processing, people may find conversations with a robot indistinguishable from talks with people," says Martin Ernst. "They may find the machines preferable."

Here we are, hardly settled into our electronic cottages, and already the age of electronic pals is upon us. Not only will our homes be electronically sentient, ready to respond to our fancies, to satisfy our whims, but inside will be a mobile entity, crammed with microchips and programmed good will, our affectionate aluminum sidekick. Get ready for bumper stickers saying, "Have you hugged your robot today?"

The Automated Society
Robert U. Ayres

Since robots are not on the verge of taking over all jobs—or even a very large fraction of them—there is no question of humans becoming parasites living off the efforts of hordes of robotic slaves. . . . Robots will play an important, but definitely subsidiary, role in the industrial economy of the twenty-first century, in much the same sense that gasoline engines play an important but subsidiary role today.

There are a little fewer than 30,000 of them—industrial workers that can be said to be true robots—all over the world. Of them, 6,000 of them are in the United States, where they are still a working minority. For every robot in the U.S., there are 40,000 people. For every robot working in factories, there are 1,000 human workers; and for every robot doing the truly dull, routine, repetitive jobs that robots do best, there are perhaps 400 humans doing the same thing. These machines are not glamorous to behold. For the most part they are immobile and insensate. Little more than a stiff, ponderous mechanical arm with one elbow, one wrist, and two fingers, each of these machines does a simple job—spot welding, spray-painting, unloading die-casting machines—and does it strictly from memory. These are the facts.

Perceptions are also facts, of a sort. And what perceptions tell us is that this small cadre of working machines represents something significant. Media coverage of robotics over the past few years has been frenzied, to say the least. Each year there have probably been more articles and interviews published on the subject than there are robots sold. Pundits see robotics and flexible automation as the salvation of an aging, sclerotic U.S. industry that is being soundly thrashed in the marketplace by the Japanese. Investors are now seeking hot new robotics stocks (sometimes in vain, as seekers far outnumber available stocks). Upscale consumers seem to be waiting eagerly for the first generation of household/personal robots to take out the garbage or vacuum the floors—or just entertain them. In the midst of this frenzy, many factory workers think about robotics, too, worrying that it will cost them their jobs.

To be sure, robots like *Star Wars'* C3PO still belong to fantasy— for now. Still, the question constantly arises: Is today's science fiction tomorrow's reality? In science fiction the convention is that while robots are stronger and more durable, humans always maintain dominance or regain their dominance at the end of the book or the movie. But what about the real world? Do real-life robots have the potential of eventually taking over most—perhaps all—of our jobs?

Or are we overreacting, endowing robots with too much potential? After all, we are the product of 500 million years of biological evolution, not to mention 5 million years of social evolution. Granted, machines are stronger and tougher than humans. Granted, computers can do specialized numerical manipulations much faster than we. And granted, computers can acquire knowledge from other machines faster than we can. But we still have inborn abilities that would be very hard—if not impossible—for machines to imitate.

Even among animals, humans are outstanding because we are the least specialized, that is, the most flexible and adaptable. The human hand, to offer one example, lacks the specialized sharp claws of a predator, the spatulate claws of a digger, the webbing of a swimmer, the feathered surface of a flyer, or the hard, horny hoof structure of the ungulates. What the hand is, is an ingenious, general-purpose manipulator.

Similarly, our heavy reliance on vision gives us one of the best general sense adaptations. Usually nature has to choose among alternatives, and it can take different forms. Many ground animals primarily use smell to deal with their worlds. It is an excellent way to track prey or find a mate, even in the dark, but it is relatively ineffective in detecting enemies approaching from downwind and quite useless for tasks that require some manip-

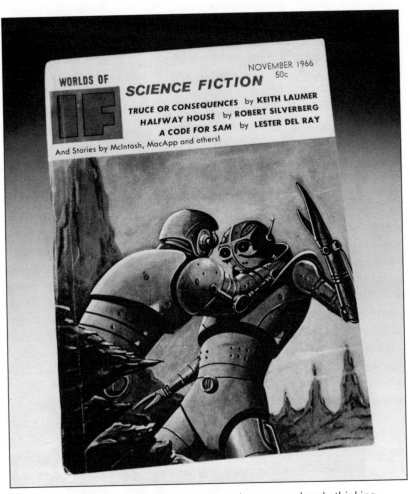

In the glory days of science fiction, writers were already thinking about the implications of a society populated by smart machines.

ulation. It is impossible to walk with your head held up to look around and simultaneously sniff the ground. Bats and the most advanced marine animals rely on another sense, hearing—bats because they fly at night and live in caves, the sea mammals because water conducts sound much better than light. But for most land animals who function in daylight, binocular, color vision seems to be the most adaptable, general-purpose sensory system. While nocturnal animals with good night vision seek dark crannies to hide in during the day, humans can operate in both daylight and darkness.

Even for us, our adaptability has its limitations. The qualities that make a good jockey are utterly wrong for a basketball player. Poets and artists generally do not make good accountants. Surgeons do not box, and boxers wouldn't make good surgeons. The point is that neither humans nor machines can be very good at everything at the same time. It is and most likely always will be a fact of life that there will always be ways in which we will be superior to machines, and machines superior to us.

How do machines surpass humans? Mechanically, they have several advantages. First, machines can be made very strong (if enough power is available). Second, their metal skins make them relatively impervious to surface damage. Third, their inherent rigidity makes it possible to repeat a sequence of motions many times with high precision and minimal wear. Fourth, they can perform motions that human joints do not permit. Fifth, they can continue to function for long periods as long as power is available, impervious to fatigue or boredom.

These mechanical advantages arise from one central fact: Most machines are made of dense, hard, rigid materials—mainly metal. But machines also have some significant disadvantages arising from the same central fact. One is that when most machines break down, they tend to be *totally* nonfunctional until they are fixed. This includes the period during which they are being repaired. Broken machines do not fix themselves; they do not heal, as living organisms can often do. Another disadvantage of machines is that their strength and rigidity can make them inappropriate for tasks involving close contacts with brittle or easily injured materials.

But let's turn the question around and ask: How do humans surpass machines? Humans seldom memorize a sequence of motions completely. In practice, we rely on visual or tactile senses to control fine motions (as opposed to gross movement). Moreover, it seems that this constant sensory feedback is really essential, even in many routine tasks. Machines programmed to repeat a sequence of motions from memory alone cannot reliably perform insertion tasks—engaging a nut on a threaded bolt, for example. This is a very hard task for a machine lacking tactile, or force-feedback, sensory capability. Often the nut will jam, and the machine, not knowing any better, will keep tightening, stripping the thread or causing even worse damage. (This is why auto mechanics often use mechanical bolt tighteners only after they have properly engaged the nut on the bolt by hand.)

Most people can immediately think of many everyday tasks that would offer comparable difficulties for insensate machines. In their present state of development, machines find it very difficult to discriminate shades, colors, shapes, and textures; to manipulate soft and floppy objects; or to function in an environment that is changing rapidly and unpredictably. This is due to the inability of machines to process and interpret sensory information and to modify their actions accordingly.

While humans find this kind of task comparatively easy, we find it difficult to respond to information that cannot be easily seen or heard or sensed by touch. For instance, our inability to sense radar signals makes it likely that computers linked to radar systems will totally replace human air traffic controllers in the not very distant future. Humans also find it very difficult to discriminate or manipulate very small, microscopic objects; to manipulate very large, very heavy, very hot, or very cold objects; or to work in space, under water, or in a very noisy or toxic environment. Finally—and this is a very important fact indeed—humans find it difficult to perform the same task over and over again without getting bored and fatigued and making mistakes.

Summarizing what we know or can reasonably infer about the inherent capabilities of humans and robots (i.e., machines) will produce a chart something like the one on pages 242–243. Moving from left to right, the tasks become more difficult for humans. Moving down, tasks become increasingly difficult for machines, assuming they are made of metal and utilize known types of sensors and strategies for sensory information-

Pick & place a medium-size oriented metal part		.
Load/unload a machine tool		
Spot weld a repetitive pattern		
Sandblast a wall in the open		
Spray-paint a simple surface in the open	Arc weld along a seam repetitively	Assemble a 2-D object (e.g., PC board) repetitively
Drive a train on tracks	Spray-paint a complex surface repetitively in the open	Assemble a wooden cabinet, electric motor, or pump repetitively
Pick & place a randomly oriented part from a bin	Arc weld a broken metal part in the open	Inspect a PC board for faults
	Wash windows selectively	Build a brick wall
		Cut coal from a face
Pick unripe fruits/vegetables by size alone	Drive a car or truck, no traffic	Build a stone wall
		Operate a tractor
Select a part from a pile of mixed parts	Wash dishes, glassware, individually	Take off a small plane in the day, good weather, no traffic
Pick & place very floppy objects, e.g., thread, yarn, wire	Inspect eggs in a hatchery	Assemble a wire harness
	Cut meat; clean fish	Assemble & sew a standard garment
	Harvest ripe soft fruits/vegetables	Inspect schoolchildren for cleanliness
		Inspect soldiers for neatness
Cut flowers individually for sale	Separate crabmeat from shells; clean shrimp	Inspect seedlings in a nursery for quality
	Plant rice or similar seedlings	
		Bathe & dress a baby or invalid

Pick & place a heavy metal part		Pick & place a very heavy metal part in a hot, noisy (toxic) environment	Laser brain surgery
Spot weld in a noisy, enclosed space	Solder very tiny wire connectors repetitively		
Sandblast on a scaffold			
Spray-paint a simple surface in an enclosed space	Operate a fire extinguisher inside a burning building		
Assemble a typewriter repetitively	Spray-paint a complex surface repetitively in an enclosed space	Assemble a large turbine repetitively	Land a spacecraft in good weather
	Machine a complex part to order	Assemble a mechanical watch repetitively	
Blow glass to order	Arc weld a broken part in an enclosed space	Arc weld a broken part under water at night	Explore the surface of Mars
Carve wood to order			Weld a broken water-pipe from inside
Cut marble or granite to order			
Finish (e.g., lacquer) a cabinet to order	Inspect an LSI chip for faults	Cut a gem to order	Arc weld a broken part under water in a storm
	Identify counterfeit paper money		
Land a small plane in the day, good weather, no traffic			
Cut & assemble a suit to order	Land an airliner in the day, good weather, with traffic control	Control air traffic at a busy airport	Inspect a VLSI chip for faults
	"Invisibly mend" a garment	Land an airliner at night in bad weather	
Drive a truck or bus through traffic	Shear a live, wriggling sheep	Identify a counterfeit "old master"	
		Diagnose a medical condition	
Deliver a normal baby & inspect for faults	Routine dentistry (cavity drilling & filling)	Repair a damaged "old master"	Land a military jet at night in bad weather on an aircraft carrier
Dental hygiene	Routine surgery (e.g., appendix removal)	Plastic surgery	
		High-speed auto chase through city traffic	Heart or liver transplant

processing and decision-making. Clearly, if totally new physical principles are discovered and applied, these relationships may change; but for the next few decades, at least, this picture seems likely to remain valid.

Generally speaking, the top left-hand corner of the chart includes tasks that are fairly easy for both humans and machines; the lower right corner includes tasks (doing a heart transplant, landing on an aircraft carrier) that would be difficult for either—and at present are the exclusive domain of humans; the upper right corner includes tasks that are likely to be easier for machines. Tasks in any *row* should be of (roughly) equal difficulty for a robot, while tasks in any *column* should be of equal difficulty for a human.

Generally, giving robots increasing sensory feedback will enable them to gradually take over tasks starting from the top right corner and working down toward the bottom left; this "rule" is obviously subject to economic and other constraints. For example, while most aircraft landings and takeoffs are quite routine and therefore suitable tasks for robots, humans are likely to be more competent at handling emergencies for a long time to come.

Tasks that are intrinsically nonroutine—for example, firefighting or mechanical repairs—are also less likely to justify the use of sophisticated robots than frequently repeated tasks. The economics of substituting machines for humans are also strongly influenced by the conditions under which a human would have to work. Human workers are naturally reluctant to take or keep boring, unpleasant, and dangerous jobs. To attract workers for these tasks, employers must pay higher wages. This favors early use of robots for such jobs.

Clearly, sensory feedback is an essential requisite for many tasks. Outside the factory, high-level sensory capabilities are required for virtually all tasks, with rare exceptions. Inside the factory, however, it is sometimes possible to engineer more predictability into a task by careful engineering of the entire production system. Flexibility of product design offers further potential for reducing the need for sensory information. A given product (say an electric mixer) may be assembled by any one of a variety of sequences of operations; today's product designer can choose designs adapted particularly to machine assembly.

Thus, if one version of the product involves a task that is intrinsically difficult for a machine, it may be possible to eliminate that task by redesigning the product to use a simpler component. Another way to minimize assembly difficulty is to minimize the number of hard-to-handle parts, by eliminating screws, bolts, washers, and nuts.

During the next two decades, the major uses of robots (and robot vision) will undoubtedly be in factories. More and more individual machines are linked together via central control networks into clusters and clusters of clusters. In the past these linkages have been strictly mechanical, but in the future computers offer an alternative link. In effect, computers will become the foremen, gradually taking over the monitoring, coordination, and scheduling of jobs that human workers have hitherto performed. In this context, the role of robots is primarily materials-handling, i.e., passing workpieces from a machine to an inspection device and on to the next machine. Robots also can and do handle tools in situations where the workpieces are large, or the tools are small, and it's more convenient to move the tool to the workpiece. They now routinely do welding and spray-painting, but they can also manipulate portable routers, electric drills, jigsaws, and glue guns.

Electronically integrated systems are tremendously flexible. They can accommodate design changes by reprogramming and can achieve high production rates without the costs and complications involved in redesigning conventional production lines. While it is not yet clear if electronically integrated technologies are economical for very large-scale production—because flexibility requires greater complexity—it is clear that integration can bring significant savings to batch producers who make a limited number of specific products.

Another factor that will affect batch production more dramatically will be the integration of computer-aided design (CAD) and computer-aided manufacturing (CAM), jointly known as CAD/CAM. In principle, we can design, redesign, and test new machines and goods as computer graphics models constructed entirely inside a computer. When we have an acceptable design, we can then instruct a team of computer-controlled machines to build what has been designed on the computer screen. For manufacturing goods in small quantities, like airplanes, which

Machines can produce creations like Cyberman,
a computer-generated manikin used by Chrysler designers
to test car-interior designs.

Anticipating the day when computers would generate CAD/CAM drawings,
a mechanical artist/automaton sketched this ship.

are very expensive to produce, this is an excellent technology. Not only does it save on design costs—cutting them by as much as half—it also saves time; the computer does many of the repetitive computations required in engineering design and does the work of the draftsmen as well.

CAD offers other advantages. First, it reduces the chance of human error, the source of many of the flaws that plagued designs in the auto industry, for example. Second, it allows more variations of a design to be tested. As a result, the finished design can be improved significantly. Third, the design process can be speeded up enormously. Once a CAD system is in full operation, new products can be generated in a fraction of the time formerly required, thus abbreviating the time needed to bring a new product (or a new model of an older product) to market.

Fourth, and finally, CAD automatically generates parts lists, specifications, and blueprints, which can then be sent electronically to the factory. The ultimate objective of manufacturing engineers is to integrate CAD and CAM, that is, to send computerized part specifications directly to machine tools. This is already virtually state-of-the-art for one special category of manufacturing: cutting or engraving flat parts (for later assembly) such as sheet metal, paperboard, plywood, printed circuit boards, computer drips, and fabric.

It is increasingly clear that the only viable long-term competitive strategy for U.S.-based manufacturing firms is to shift quickly to flexible batch production and at the same time to accelerate their traditional rate of product innovation. Quality and performance must displace price as the major objective of product design. Only by this strategy can they counteract the competitive advantages currently enjoyed by Japanese manufacturers. Obviously, a competitive strategy based on rapid innovation for U.S. capital goods producers depends on rapid adoption of CAD/CAM and robotics.

How will a newly built factory in the year 2000 differ from one built in the recent past? Taking into consideration all of the points mentioned above, the differences will be quite profound.

First, the local availability of cheap labor will not be a major consideration because the plant will need very few people, even during the day shift. Second, it will operate virtually unmanned at night and over weekends. Third, computer control will sharply

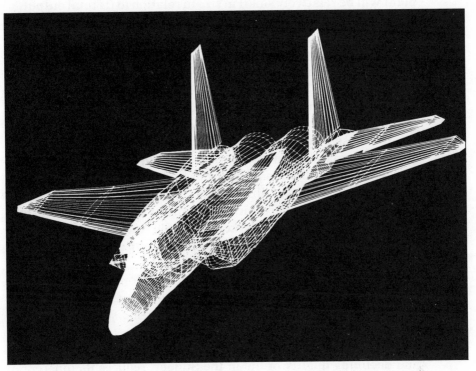

Precise CAD/CAM drawings like this help manufacturers anticipate costly
problems when building multimillion-dollar aircraft.

reduce the need for on-site storage of raw materials and semi-finished parts. Fourth, for the reasons mentioned above, the new plant will be relatively small in size in relation to its total output, as compared to present-day plants of comparable capacity. But for its size it will generate much more goods traffic than a present-day plant.

The typical, enormous, 1960s plant spread over several hundred acres, surrounded by vast parking lots, and employing armies of hourly paid workers will become obsolete. The flexible factory of the year 2000 will still be located close to major rail, truck, or air transportation facilities, but it will not itself require a great deal of land nor many workers. It may, therefore, be closer to the city center than most factories built in the '60s—probably near the major freeway linking the city center with its airport. Its location will be determined primarily by transportation costs and closeness to major markets. Flexible automated factories will not have to be located in remote rural areas or foreign countries simply to minimize labor costs.

Thus the impetus that has shifted so much production away from major U.S. population centers in the Northeast and Midwest to the Sunbelt will eventually exhaust itself. Sometime around the year 2000 the pendulum will begin to swing in the other direction as the congestion, increasing water shortages, and declining quality of life in the Sunbelt reduces its inherent attractiveness.

What of the 14 million or so hourly paid blue-collar workers in the United States? The forces of international economic competition from the Far East, not the threat of competition by robots and "steel-collar" workers, are already rapidly reducing their ranks. With each downswing of the economic cycle, hundreds of thousands of highly paid steelworkers, autoworkers, and their unionized brethren are now being laid off. With each upswing a much smaller number are rehired. Employers are closing their oldest, most unionized plants and shifting production to the South or overseas. Unions are becoming weaker, and the general public (outside of a few states where their influence is still strong) is not impressed by their arguments for protection—if it means higher prices for consumers.

Unless this situation changes dramatically (which I do not

expect), the major industrial unions such as the UAW and the USW will be severely weakened by the year 2000. Many of the younger factory workers are already recognizing the inherent insecurity of their present jobs and are beginning to seek education and training to fit them for other careers. The older workers, by and large, will continue until they retire (early, in many cases), but they will not be replaced. Cities like Pittsburgh, Detroit, and Chicago are already beginning to show the effects of the coming changes.

There will, predictably, be some casualties. There are a number of smaller cities and milltowns with only one major plant; many of these plants are already obsolete and likely to be shut down rather than renovated or replaced. Some of the older workers with seniority (especially steelworkers) are protected by their union contracts and will simply retire. Others are less fortunate; some will lose their homes and be impoverished, unless significant government-sponsored relief efforts are undertaken to help them relocate and retrain. Many will be forced by circumstances to take jobs paying much lower wages than they have been accustomed to receiving.

The most plausible method of providing assistance for displaced industrial workers in the smaller cities and milltowns is to embark on some major (but limited) regional renewal projects. Former steelworkers and autoworkers could be employed, with minimal retraining, in construction jobs. The challenge will be to identify and finance appropriate large-scale construction projects, manage them well, and ensure that the newly created jobs are not monopolized and that displaced industrial workers have a fair chance at the resulting job openings. This almost certainly requires a public-private "social contract"—similar in concept to the Municipal Assistance Corporation that pulled New York City back from the brink of bankruptcy a decade ago.

Although the natural domain of robots is gradually shifting from the upper right toward the lower left of the chart on pages 242–243, it is by no means certain that this shift will extend into other areas of the diagonal. One might conclude differently if technological progress were a truly autonomous phenomenon, independent of socioeconomic driving forces. In effect, this trend— if it exists—implies robots and computers are a kind of evolving species of life, or at least pseudo-life. This is a fascinating conjecture, but it makes very little real-world economic sense. The answer goes to the core of human-machine relationships.

In brief, humans are a species with an extreme degree of flexibility and adaptability to change. We humans have become specialized in being unspecialized. We can learn to do almost anything, in time. But, because we do not learn particularly efficiently, it is hardly surprising that our inborn capabilities are often easily surpassed by special-purpose machines, or even another animal species. It follows that we humans tend to use tools and machines to extend our limited inherent capabilities—*but it also follows that tools and machines are invariably much more specialized than we ourselves are.*

This is a point that sometimes causes confusion and needs to be explained. Obviously, some machines and factories are much more specialized than others. An automobile engine plant is one extreme: It is totally specialized to make a single product. By comparison, a general-purpose machine shop is relatively unspecialized, meaning that it can be used to custom-produce a wide variety of different sizes and shapes of metal products. True, the auto engine plant produces engines much more cheaply than the machine shop, but by sacrificing the ability to make anything else. The trade-off between flexibility and efficiency is quite a general one—both in nature and in engineering. Mass-production economies are achieved by exploiting this relationship, accepting rigidity and inflexibility as the price of maximum efficiency. But with a commitment to mass production comes the inability to respond quickly to changing markets.

That some machines such as robots can be much less specialized than other machines must not obscure the fact that even the *least* specialized machine cannot begin to compare with humans. There is no such thing, really, as a general-purpose machine. One speaks of general-purpose shops or general-purpose computers, to be sure, but this is a semantic trap for the unwary. True, some computers (for example, analog devices) are more specialized than others, but the range from most flexible to least flexible is narrow. By analogy, one might say that a twelve-string guitar is less limited than a six-string guitar, but both are far more limited than a full orchestra. The comparison between machines and humans is like this comparison between guitars and orchestras. The orchestra is not simply a collection of instruments: It alters its composition and capabilities for each piece of music. It may feature a guitar for one piece, a harp for another, and an organ for a third. So it is with humans. Not only are humans capable of a wide variety of actions, but we uniquely have the ability to use specialized tools (that is, machines) to solve particular problems efficiently, as needed, without sacrificing their own inherent flexibility.

Seen in this light, humans and machines are complementary, not competitive. While it might conceivably be possible for robots to be given humanoid form (two arms, two legs, eyes, voice, etc.) and electronic brains capable of imitating some human thought processes, there would be no valid reason for ever doing so. The result would be an unnecessarily complex, unreliable creation, not very good at any specified task and enormously costly. The simple point is that no single robot would ever be required to perform the whole range of jobs a human can do, even in the unlikely event such a robot could do them all better. To say it another way, individual robots will always be designed for specific tasks; hence, robots will always be specialized. Why give robot arms more degrees of freedom than the task requires? Why make a robot bigger, stronger, faster, or smarter than necessary?

Industrial robots are rapidly differentiating into many subspecies, all with a wide range of diverse capabilities (size and weight, speed, precision, number of arms, number of fingers, number of joints, sensory capabilities, etc.). The resemblance between a welding robot and an assembly robot is already slight, and they're likely to become even more different in the future. Mining robots, undersea exploration and repair robots, and space construction robots will be even more specialized. The evolutionary trend of industrial robots is clearly away from the general-purpose humanoid form, not toward it.

Before considering the long-run implications of this, we should perhaps look at it in some other context than the factory. After all, as I have pointed out, the factory environment is inherently more controllable than the general human environment. It is less demanding of high-quality sensory capabilities and intelligence. Would the ability to operate in other, more variable environments (such as mines, highways, farms, households, or battlefields) justify developing a humanoid robot?

As soon as the question is posed this way, the answer seems clear. Each of these uses for a robot is itself specialized. An undersea robot would have to have very different capabilities and characteristics from a construction robot, a household robot, or a robot soldier. Thus, the undersea robot would be an autonomous unmanned submarine with one or two manipulator arms and some sensory devices based on sonar rather than vision, all adapted to the marine environment. The mining robot would probably be an armored, tracked vehicle specialized in doing a specific task such as drilling holes and inserting roof bolts. The farming robot might be an autonomous, intelligent tractor capable of orienting itself at all times to the boundaries of the field while towing various pieces of equipment. (But it would probably have to call for help if it got stuck in the mud.)

Like the mining machine, the robot soldier would also probably be an armored, autonomous vehicle, but with no particular top or bottom. It might come in several variants for different kinds of terrain, ranging from spheres, air-cushioned vehicles, and wormlike crawlers for flat or marshy ground to insectlike multilegged forms capable of maneuvering through forests.

The robot truck or bus driver would be, in reality, a robot vehicle *without* a driver. The problem of designing a general-purpose, humanoid robot to drive a vehicle safely in traffic is not worth solving, if it could be solved at all. When driverless vehicles are ultimately developed, they will probably be controlled by a network of sensors and computers implanted in the road system. The passenger would simply specify a final destination on a display map, and the vehicle would communicate these data to a regional traffic-monitoring computer, which would route the vehicle and turn control over to a sectoral control computer.

The construction robot of the future would also be specialized. There may someday be autonomous bricklaying robots, concrete-pouring robots, riveting robots, welding robots, painting robots, and so on. But none of these machines will be humanoid and none will have capabilities beyond those actually required for the task. For example, a robot for construction of solar power satellites in orbit would, of course, be specialized for different tasks from a ground-based construction robot, but unquestionably it would be specialized.

Finally, what about the domestic robot? In this case, it could be argued that robots might be more "acceptable" if they looked somewhat like human servants. I concede this question to the marketing experts, but it seems doubtful that customers would demand a humanoid appearance if it added greatly to the cost while detracting from performance. The first household robot seems very likely to be a semiautomated vacuum cleaner, which, in robot mode, would move automatically around a room, sensing and avoiding obstructions, but otherwise traveling at random.

A vacuum cleaner robot would resemble a conventional tank-type vacuum cleaner on wheels, but with a shorter arm ending in an air intake. It will probably use an ultrasonic directional range-finder (similar to the devices found on self-focusing cameras) and tactile sensors on the sides and tip of the vacuum arm, which would wag back and forth automatically (somewhat like a dog's tail). In manual mode, the unit would have an extendable hose for emptying ash trays, cleaning venetian blinds and sofa cushions, etc. The robot vacuum cleaner would require a steerable driving wheel in addition to the vacuum arm. A fairly sophisticated microprocessor would be required to interpret the various possible signals from these sensors and operate the joint controls.

A laboratory prototype could probably be designed and built today, but it would cost a million dollars or so and would have to be at the very edge of the state of the art. Conceivably, such a unit could be commercialized in a decade if a large firm made a major effort, but the final product would be quite expensive even in large-scale production. And how many people would pay $1,000 for a semiautomated vacuum cleaner?

Other types of household robots are obviously feasible in principle. One of the simplest robots is an automated drink-server. The device would be an arm capable of pouring drinks from a designated set of bottles in a fixed rack and adding ice and mixers on demand. It could be programmed by means of an ultrasonic remote-control device (similar to a TV control), by a light pointer, or conceivably by voice instruction.

A simple button-operated drink server has already been demonstrated by Unimation Inc., and several other robot manufacturers have their own demonstration versions. A practical commercial version, designed for serving at cocktail parties, would probably have to be mobile (like a robot mail delivery cart). Most likely it would slowly follow a specified route, marked by a wire in the floor, announcing its presence verbally by "beeping" or by a programmable phrase such as, "Would you care for a drink or a canape?" It would also specify the appropriate response—for instance, "If you would like a drink, please touch the white button"—and would then proceed to offer a menu of its available refreshments.

This sort of robot could be manufactured today, in moderate quantity, probably for a few thousand dollars each. On really large-scale production, the price might come down to $500 to $1,000. But it's by no means clear how many people would pay that sort of price for an automated drink server of such marked limitations.

But, there would be a much greater potential market for a mobile household robot with the more generalized capabilities of "pickup," "putaway," and "cleanup." Suppose we consider the requirements for such a robot. In pickup mode it should be able to pick up and store, for later putaway, all objects not belonging in a given room. This obviously requires the robot to possess a detailed 3-D map of each room and its contents stored in its memory. It must also have a high-quality vision and tactile system to compare the actual disposition of objects in the room with the correct disposition. It must be able to accurately distinguish and ignore humans and pet gerbils, possibly by means of a CO_2 sensor combined with an infrared sensor. Apart from these feats of memory and recognition, the robot must have a rational navigation strategy based on a knowledge of its own position and orientation in the room, relative to all potential obstructions.

For each object encountered in a search pattern the robot must identify it precisely (how big is it? is it soft? hard? is it breakable? does it belong in this room?) and then determine its position. Having made these decisions, the robot must devise a pickup strategy, which is comprised of many elements (the position of the gripper, the amount of gripping force needed, how to detect and adjust for slippage, the trajectory of the pickup arm, choice of storage location, position of the object in its place of storage, etc.). Trivial as these decisions are to a human, they are far

Part of the robot mythology, films like *Forbidden Planet* suggest what life with intelligent helpmates will be like.

beyond the current state of the art of "intelligent" computers. But there seems to be no fundamental reason that computers— even small ones—should not eventually be able to solve such problems fairly efficiently.

The putaway mode would require much the same set of capabilities as the pickup mode, except that the robot must cope with the fact that objects found out of place in one room must belong in another. Having completed pickup in a given room, the robot would immediately put away those objects belonging in the same room by referring to its internal map of that room. To deal with an object belonging elsewhere, it would need a more complex strategy. It might have to call up from memory its maps of other rooms, compare the item with each map, then go to each room to find where the found object corresponds to a missing object. Or it might first identify each found object, locate it on a master list, and then look up its correct location.

For scrubbing, waxing, or polishing, the most effective motions would probably be taught to the robot in manual (or "teach") mode by a human, much as a robot arc welder is taught by an expert human welder. A cleanup robot would need various attachments, such as rotating brushes, one or more tanks for storing cleaning liquids, and a vacuum system capable of collecting liquid droplets as well as dust. To be truly automated, it would also need to have the ability to empty and refill its storage tanks, to wash out its internal plumbing with the help of suitable nonflammable solvents (which can be separately stored and recycled), and to change its own attachments.

It is likely, on closer consideration, that pickup/putaway could not be combined in a single unit with a cleanup robot, though they might be combined with a serve unit. The cleanup robot's tanks, pumps, hoses, and tool attachments would be useless (and in the way) for the other modes of operation. On the other hand, the pickup/putaway/serve robot would need only a flexible arm (perhaps two) and a flexible internal storage space for a variety of objects. (It remains to be seen whether other functions can be combined efficiently.)

What is quite clear, however, is that future robots will be specialized for particular tasks or types of tasks, and will not be endowed with a high level of general intelligence. For instance, the programming needed to enable a robot pickup/putaway/serve unit to function effectively would require the ability to solve certain limited classes of problems, such as: What class of object is x? Where does x belong? How should object x be gripped to avoid breaking it? How should object x be oriented? How should internal storage space be loaded to maintain balance? What is the best route from point A to point B to avoid obstructions? These problems are intrinsically very difficult and will require far more computer power and more sophisticated software than is presently available. There's no doubt that such a robot would be "intelligent," as the term is used by computer scientists. But robots will not resemble humans, and their intelligence will be narrowly limited to the specialized purposes for which they are designed.

When the future of robotics is discussed nowadays outside scientific and engineering circles, two kinds of questions arise with particular urgency. The first is: What are the implications of microprocessors and robotics for the future of work, in terms of the future relationship of humans and machines in the workplace? The second major question, the exact formulation of which depends on the answer to the first, is: What do the coming changes in the workplace portend for social and economic institutions?

As an instance of the first type of question, some social scientists have worried about whether humans will have to work "with" robots on future assembly lines and, if so, whether this would increase the pressures of pacing humans to machines, and the resulting alienation of human workers. As an instance of the second type of question, some social scientists are worried about the effects of a decline of the highly paid, semiskilled job level that serves as a bridge between the low paid, unskilled, entry level and the elite upper level of professionals and executives. In short, some have even foreseen the increasing substitution of robots for factory workers as a somewhat malign social force undermining the social basis of our political democracy. As a matter of fact, these basic questions have been discussed for at least a century in the context of increasing automation, and they will not be easily settled. I can only hope to offer a very modest contribution to the debate. What follows is strictly my personal opinion, based on the facts and arguments presented earlier in this chapter.

As to the future relationship of humans and the workplace, I

A scene from a stage version of *R.U.R.* The play's apocalyptic
vision still colors many of our attitudes toward automation.

believe that robots will certainly be more widely used in the future, but that the number of existing jobs actually displaced by robots over the next half century or so will be relatively moderate. As I stated already, by far the largest job loss will occur in the manufacturing industries where, by the year 2035 or so, up to 90 percent of all jobs on the factory floor will be eliminated by automation; the total number of such jobs in the United States today is only about 14 million, however. In the nonmanufacturing sectors of the economy, there will also eventually be a large number of robot hall sweepers, window cleaners, garbage collectors, street cleaners, bridge painters, farm tractors, pothole fillers, bricklayers, gasoline (or methanol) pumps, fast-food dispensers, ticket sellers, and so forth. But the net displacement from all of these specialized applications will probably be less than in the manufacturing sectors alone.

To summarize the job situation, machines will eventually take over most of the repetitive, boring, demeaning jobs in our society—especially the jobs where workers interact mainly with machines or materials, not people. But robots will *not* take over the jobs that require high-quality sensory capability, rapid responses to a changing environment, or effective interaction with other humans. In fact, most jobs are already in the latter category rather than the former. Thus by 2035 most factories will be largely unmanned except for a few security guards, maintenance personnel, and engineers. But most other places of work such as schools, hospitals, offices, shops, and restaurants will continue to be operated by humans. Similarly, mining, construction, transportation, communications, and agriculture will continue to use human workers, in most of the same functions that such workers currently perform. And handicrafts and popular arts are likely to be a major growth sector as factory employment declines.

I think, in short, that the problem of worker alienation by asembly-line jobs will be eliminated in the simplest possible way: By 2035 humans will continue to design, install, modify, maintain, and repair machines; but in the advanced industrial societies they will no longer "feed" them or directly control them. The future human worker will often be assisted by a machine, but the more unnatural situation where humans, in effect, had to help the machines because they were too primitive will have disappeared.

With regard to the second great social question, I have already noted that there will inevitably be some forseeable effects of technological progress in automation in the coming two decades. A human society should be prepared to ameliorate the economic impact of job displacements and assist the workers to retrain and relocate, when necessary.

But what of the impact on our social structure? Will the blue-collar middle class shrink or disappear? Will the opportunities for moving from entry-level positions to skilled jobs via on-the-job training be sharply restricted, resulting in an elite defined by educational opportunity? I think that there is some real danger of this happening in the near future.

The kinds of skills that will qualify humans rather than machines for jobs in the future primarily either are inborn or are learned at home or in school, but not—by and large—on the job. Eyesight, hearing, and motor-coordination are examples of inborn characteristics in humans that are not significantly improved by extensive on-the-job training. (It does not take very long to train drivers or typists, for instance, as long as they can read reasonably well.)

By comparison, reading, writing, grammar and speech skills, and logical (computer-programming and mathematical) skills are mostly learned in school. These, in turn, are essential prerequisites for accumulating more specialized knowledge. Once they have the communication skills and logical skills, people can acquire the substantive knowledge which, along with social skills, qualifies them for better-paying jobs in today's society. It is the lack of these learned skills, primarily, that holds back displaced factory workers.

The answer for displaced workers, in the long run, is not retraining (for what?) but reeducation. The middle class currently includes some blue-collar workers who would not be as well paid as they actually are if it were not for the strength of their unions. (These workers, if they lose their present jobs, are likely to slip down in the socioeconomic hierarchy.) But there are several categories of jobs that have always been poorly paid, in many cases because the jobs in question have been held primarily by women—teachers, nurses, and secretaries. Also, some service jobs will gain in prestige as the public gradually learns to distinguish good from poor service. A noteworthy example of a skilled profession that is grossly undervalued in the United States (but not in Europe) is that of waiter.

To summarize, there will probably be a change in the composition of the middle class but no major decline in its long-run importance. Formal education will probably continue to grow in importance, as the value of on-the-job training declines. Adult education for the undereducated and educational sabbaticals for obsolescent professionals will be much more common.

A final comment: Since robots are not on the verge of taking over all jobs—or even a very large fraction of them—there is no question of humans becoming, in effect, parasites or living off the efforts of hordes of robotic "slaves." This is purely a literary fantasy. Robots will play an important, but definitely subsidiary, role in the industrial economy of the twenty-first century, in much the same sense that gasoline engines play an important but subsidiary role today.

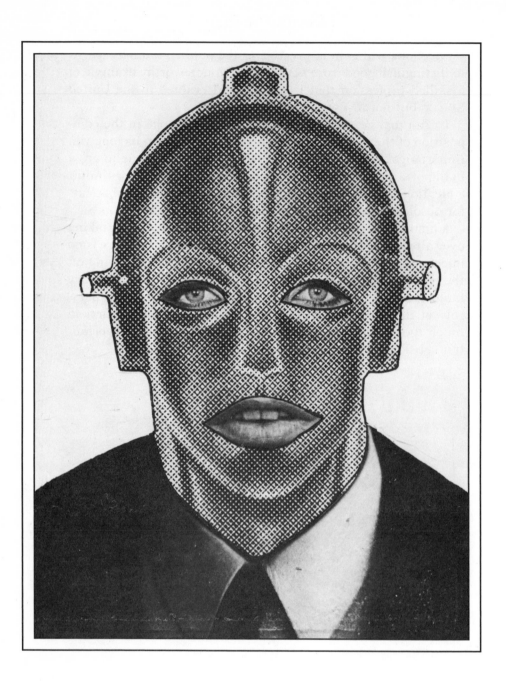

Scenes from the Twenty-first Century
Robert Sheckley

It was ironic that the first intelligent aliens encountered by mankind came, not out of the sky in spaceships, but out of our own laboratories and factories. Where we had tried to create servants, we instead had found friends.

Introduction

We've been waiting a long time for the age of the robots. Down the ages we've listened to the old stories—Pygmalion and Galatea, Hephaestus and Talus, Daedulus the artificer, Prometheus the lifegiver, the gloomy golems of Prague, Faust's mystical homunculus, and many others. More recently, science fiction has whetted our appetites for more wonders. And now, quite suddenly it seems, the time has come, the wonders have arrived, our robots are here.

Taking over from the literary men and the mythologists, the scientists are arranging this for us. Their robots work. They are getting smarter and more versatile each year, and there are more and more of them. They do everything but think. Soon, perhaps, they will do that, too.

What will a thinking robot be like? Will it possess consciousness? Will it obey Asimov's Three Laws of Robotics? Will it have an ego? A sense of self-preservation? Will it have feelings? If so, how will it feel about us? And what will happen if it doesn't like us? These are a few of the things we might consider as we steam full speed ahead into the automatized future.

A robot is man's latest way of tricking the environment into doing something for him that he doesn't want to do for himself. Robots are pure fools for work. They're going to take a lot of the routine stuff off our shoulders. They will free people from the assembly lines, the vegetable fields, and the fruit orchards, thus giving these people a better chance of becoming jet pilots, gym teachers, molecular biologists, magazine illustrators, and the like . . . if economic conditions permit.

If they don't, then we may have a problem finding something for all those people to do. Perhaps we need a Natural Intelligence Lab to develop "better" human beings, individuals who are more docile and longer-lasting, to compete with the robots. But that's in the long run. For the short run, we might consider paying the surplus people salaries to let them buy the cars, washing machines, and oven-ready chickens that our robots produce—it would complete the economic cycle. If this seems impractical, perhaps the excess people will simply fade away when they see they're not wanted. We don't even need them as buyers, really. There's no reason that robots couldn't be programmed to shop in our supermarkets, watch our TV shows, and take our cars out for Sunday drives. If necessary, the entire Earth could be put on automatic, without any humans at all (an option that is neither recommended nor likely).

The social implications of the robotic revolution are unpredictable and far-reaching. We have the blueprints for the automatic factories, but we don't have the game plans for the human lives that will be centered on them. One reason for the inscrutability of this future is the fact that we are part of the phenomenon we are trying to study. When we try to look at ourselves we get in our own way. Prediction would be a lot easier if our demands were shaped mainly by our needs, but that's not the case. We are actuated by desires and fantasies. To know how well a given product will sell, a manufacturer must guess at how well it will be received (i.e., how well it will "play"—show business being a key factor in the selling game).

What follows are some speculations on the future of robots and humans. I don't think any of these scenarios will happen precisely the way I have described them. But I'd be surprised if aspects of these stories didn't prove, in some measure, prophetic.

Scenario One: A Brief Survey of the Future

By the mid-twenty-first century, the trend in robotic design had turned decisively in favor of free-moving, general-purpose, humanoid robots.

Although the human form was by no means the most efficient for the variety of jobs the robot was called upon to perform, it was by far the most popular. It had what the designers called built-in acceptability. People have always enjoyed looking at other people. They liked looking at imitation people, too.

Despite the popularity of human form, the public wanted to be able to differentiate clearly between humans and humanoids. But this presented no difficulty. No serious attempt was ever made to solve the many technical complications involved in simulating the human repertoire of movements, expressions, sounds, and gestures. No one tried to construct in a robot brain a sensorium capable of recognizing and responding to all the cues and subcues that humans are forever exchanging, and that are the epitome of humanness. All of this could have been accomplished, no doubt, if mankind had been willing to put sufficient energy into what can only be called aesthetic engineering. But the design trend in robotics did not go that way. People didn't want lifelike robots. They were pleased with the current models' jerky movements and flat, artificial voices; their silly fixed smiles; their clothing that never looked right on them; and their earnest, bumbling air, which, together with their small stature (the most popular height for robots was between 3 and 4 feet) gave them a sympathetic, endearing appearance. The sight and sound of a robot was always good for a chuckle. They were the butt of many jokes. It was reassuring to know that no matter what else these creatures could do, they could never pass for one of *us*.

The first generation of free-moving robots was imperfect in many respects. Metaprogramming of common sense was still in its infancy, awaiting Drayton's invention (in 2063) of time-tagged, self-canceling propositional matrices. But in the early days, robots often fell into ambiguities and double-binds. You'd see a robot standing patiently in front of a wall at the end of a blind alley, unable to go on but unwilling to turn back and take a

" 'We have the highest regard for our relationship with humans.
But we would like to explore what it means to be a robot.' "

more circuitous route. "The map says this street goes through," the robot would explain, unable to consider that the map might be wrong. It was an act of simple kindness to question these robots, find out what their trouble was, and give them the commands that would clear up the difficulty and permit them to go on about their business. As children we used to impose on these patient humanoids no end, sending them on crazy errands, asking them meaningless riddles, begging to be taken up for robotback rides.

Some of the early intelligent robots suffered from malfunctions that defied easy analysis but presented clearcut analogies to human mental states. *Information retrieval blockage* was similar to hysteria; *biased information scanning* resembled paranoia, and *self-referential fixity* was much like catatonic withdrawal. It was difficult to determine whether these quasi-mental simulations were caused by hardware or softwear deficiencies, or a combination of the two. I remember one robot who complained of "lazy circuits" and "tired batteries" (typical robot efforts at self-analysis!).

And once I came across a robot sitting all by himself on a park bench, apparently lost in thought. He knew where he was supposed to be going, *specifically*, but claimed that he had urgent, unanswered questions about direction in a *general* sense. This disturbance over generalities resembled the human condition of angst or anomie. Apparently he had been built with sufficient capacity to allow him to become aware that he didn't know the overall purpose of his existence. Before he went on with his specific programming, he told me, he felt that it was only logical to try to determine the general conditions under which he operated. When I pointed out that the question was not computable, he replied that it was nonetheless inescapable.

Quite a few of the early intelligent robots became infected by these considerations, which came to be identified as pseudo-metaphysical spirals of logic. These were the result of man-to-machine overview imprinting, also known as *inferential programming*, and when the phenomenon was more clearly understood, as *covert conditioning*. Sometimes the condition could be cleared up by reasoning with the robot. More often, complete reprogramming was called for. Occasionally, on account of the

ghosting phenomenon (*circuit bias ghost effect imprinting*), entire
hardware had to be replaced as well.

That robots could suffer from vague but pressing problems of
a personal nature was not unwelcome news to the general public.
Many of us found it reassuring that our robots were fallible. And
perhaps we needed some reassurance. Things were moving along
more quickly than our ability to react to them. We seemed to
be caught up in some kind of contest—all of us, the whole human
race—where we didn't know who our opponent was, but we had
the uneasy feeling that we were losing. It was profoundly de-
pressing to know that a computer had finally become the chess
champion of the world. And some of us were none too pleased
by the brisk sales of computer-written books, and the gleeful
way some reviewers had of imputing to these books "insights
rarely found in a human being." (Some of us even suspected
that robots might be writing the reviews.) The runaway success
of *Robots and Love* by Tony 34452S was especially disturbing to
many people. "It's the robots who are buying it, what else would
you expect them to read?" was a typical comment of that period.
But it wasn't funny. Our perversity in preferring robot art to
our own was proof to some that if the human race should come
to a sudden, untimely end, it would have no one to blame but
itself. Unprovoked attacks against robots mounted. It was fortunate that robot-written books turned out to be a short-lived
fad.

The main impact of robotics was on the industrial-economic
scene rather than the cultural. Paradoxically, it was the trade
unions that set off the robot explosion. Alarmed at seeing more
jobs vanish each year to automation, the unions finally agreed
on the now famous Proposition 22: that an employer would be
permitted to dispense with a human worker only if the human
was replaced by a union-approved, general-purpose programmable robot whose salary (as calculated on the newly developed
Nonhuman Worker Index of Profitability) would be awarded to
the replaced worker and the union, in equal parts, in perpetuity,
or until the machine had been ruled legally nonexistent. After
considerable debate and delay, the manufacturers accepted this
formula, having determined that, even with robot salaries figured in, it was still more profitable to employ robots than to
continue with humans. For example, robots need not be covered

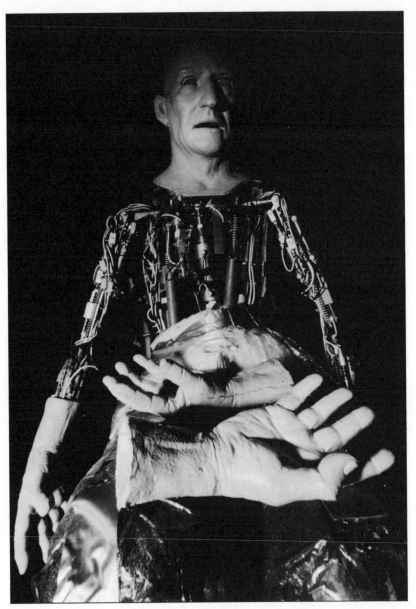

"It was reassuring to know that no matter what else these creatures could do, they could never pass for one of *us*."

by medical plans and are not awarded vacation time. Christmas parties and incentive bonuses can be omitted. The huge space previously given to workers' cafeterias, with their perpetual smell of boiled cabbage, can be turned to something more directly relating to the manufacturing process. These are not insignificant gains.

And there was the added advantage that robots do not belong to unions and therefore can never go on strike. The unions fought bitterly for the right to unionize the robots. But a key Supreme Court decision went against them. In *Larry 3442S* v. *General Motors*, it was held that a robot was not entitled to sue for overtime because he could not be said to comprehend that he was working at all, which knowledge was held indispensable to the right to make a legal plea. The key issue here was the question of introspection; the unions claimed that a robot was "an individual and aware of himself as such" and therefore entitled to union representation. Yet in *Jim Jim 35572Q* v. *Litton Industries*, the robot plaintiff was declared incompetent to plead in his own behalf "despite impressive intellectual attainments in the area of pure mathematics," because he lacked "that faculty of awareness of the ability to introspect which constitutes the basis of what we commonly call mind and recognize as the sole basis for a claim to individuality." For legal purposes, then, a robot could not be considered a person since, by existing criteria and commonly accepted methods of detection, the robot had no mind with which to consider himself. Still, the ruling hedged by adding that this must not be held to infer that such a mind might not in fact exist, or be discovered or invented at some future time.

These legal niceties over whether intelligent robots possessed anything comparable to mind, consciousness, ego, or personal identity were mostly ignored by the public. Robots were working in the new factories in ever-increasing numbers, and people's attention was captured by the visual drama that this afforded. In many American cities, crowds once gathered to watch the robots file out of their slum dormitories each morning. The robots dressed like nineteenth-century proletarians complete with heavy shoes and with cloth caps pulled down over their faces. They streamed to work with lunch pails (in fact, portable battery packs) under their arms. The robots didn't need dormitories, of course, but it was all part of the consciously staged spectacle of those early, innocent days. Although people knew better, they enjoyed thinking of robots as having real lives, needing to lie down after a hard day's work, wanting a good stiff belt (of electricity) in the evening.

Manufacturers were quick to exploit the street drama of their machines. They hired designers, stylists, and scriptwriters, and joined in the fun that was so good for business. They set up special bars and night clubs patronized exclusively by robots (until—as planned—the trendsetters and "fun" people discovered them), robot dance halls, theaters, recreation areas, even a robot swimming pool filled with glycerine!

The trend toward humanizing robots received additional impetus from the immense popularity of the TV series *The Heskas*, which told of the trials and tribulations, successes and failures, loves and hates, of a family of robots in a small town in Missouri. Complete fantasy, of course, sheer fabrication, scientifically unsound, technologically impossible—but the greatest box office success of the decade. When Lance Heska went west to become a cowboy, the network was flooded with offers of work for him. When Tansy Heska was drugged with an injection of "noxious electricity" and kidnapped in Miami Beach by a cult of yellow-skinned devil worshippers, people reported seeing her in twenty-three states and most of western Europe, and the government of China protested at the "slanderous and unwarranted characterization of the devil worshippers as Orientals." Apologies were made, sales boomed, and every child had to have a complete set of Heska dolls. These were only the first of many profitable spinoffs.

In contrast to the cheap, mass-produced industrial models, the first generation of free-moving personal-service robots (valets, ladies' maids, chefs, butlers, etc.) were extremely expensive. Owning one was a status symbol, a sign that you had arrived. Perhaps by default, America led the world in those early days of personal-servant design, because the energies of the Japanese were taken up in starship design and space colony planning. American diplomats used to bring their robot servants with them to foreign posts, where they excited much admiration and envy. The diplomats soon discovered that you could sell a robot servant in Europe for many times what you paid for it in the United

States. Even better than outright sale, you could hire out your robot to a foreign entrepreneur and collect the machine's wage, thus ensuring for yourself a long-term income. There were no laws at that time controlling the movements of robots across international borders or preventing them from competing in job markets where their superior technology and programming flexibility made them highly desirable.

Most countries were quick to recognize the advantages of being able to use these self-contained units of American technology. In the Third World, it was quickly discovered that a general-purpose robot could be used as the basis for many one-family industries, especially in such occupations as metalworking, weaving, pottery, and kebab-making. Manufacturing robots for export became one of America's chief businesses until the Europeans and the Japanese caught up.

The European robot industry developed rapidly. Europeans were quick to see that robots could be used as a form of national advertising. Congresses of Culture were set up in each country to determine their national cultural traits, and legislation was passed requiring their national robot manufacturers to simulate a minimum percentage of these traits. Many of us still remember the early German models with their lederhosen and Tyrolean hats, the Norwegian robots with their cute little sailor hats, and the Italian robots with their expressive shrugs. These trait-tagged machines were the forerunners of what came to be known as the Great Robot Immigration.

It was only natural for foreign robot manufacturers to take advantage of the free trade situation and export their robots to America to compete in our work force and remit their wages back home. The Americans, who started the practice, were nevertheless caught by surprise when shiploads of robots from all over the world started to land in our ports, each group accompanied by a bonded negotiator, typically a fast-talking individual, who negotiated their wages on behalf of the absentee owners. Although robots need no habitation, the law at that time required all machines simulating man in any respect and to any degree to live at a definite address and to dress decently. Like so many immigrants before them, the foreign robots were clustered in the worst slums. There were other restrictions as well. I can still remember the ghetto on the outskirts of my home-

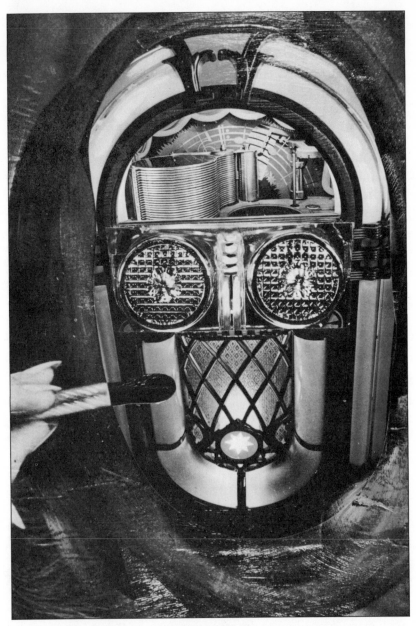

"People didn't want lifelike robots. They were pleased with
the current models' jerky movements and flat, artificial voices."

town, Maplewood, New Jersey. It was inhabited mainly by Polish and Lithuanian robots. They had bearded faces and wore long black gabardine coats. I can still remember their Old World courtesy and their incomparable folk songs.

Another development during those early years of the Robotic Revolution were the robot toys of every kind and description that dominated the market for a decade. These toys quickly became indispensable for entertaining, educating, and caring for the young. Many children preferred them to other children. They would obey Granny Gremlin (one of the top-of-the-line babysitting robots, produced by Lionel) but wouldn't listen to the teenage babysitter from across the street. And no wonder: Granny Gremlin didn't spend her babysitting time talking to her boyfriend on the telephone or watching TV. Granny Gremlin was attractive, inventive, motherly, but with a youthful outlook. Amusing a couple of kids for a day or week or forever was no problem at all. Baby Module Centers, as they were called, came in a variety of ideational programming modes—Montessori, Rogerian, Modified Hippie, etc.

The preeminent symbol of that optimistic era was beyond doubt the Cuddles Cat, one of Standard Brands' most successful patents. About the size of a standard household tabby—slinky, soft, warm, affectionate, yet independent (through the use of randomly selected hierarchies of choice)—it was capable of singing a song and telling a story, and also able to eat food and lap up liquid. The Cuddles Cat raised an entire generation of American children, and provoked the writing of many reminiscences.

The Cuddles Cat didn't actually eat. Its consumption of food was a scavenging function. The robot was programmed to trail along after the children, "eating" the crumbs, apple cores, half-consumed peanut butter and jelly sandwiches, and the like that it found scattered around the playroom.

On a more sinister note there was Mandrake, the robot killer wolverine, a product of the strange talents of Down Laboratories of Oakland, California. Advertised as the pet for people who hated pets, the robot killer wolverine pretended not to understand anything and never spoke a word. It only grunted or snarled, and it refused to obey even the simplest commands unless they were shouted at it. The beady-eyed little machine defecated small, soft, aromatic orange masses at random intervals; it snored loudly when it "slept," attacked and savaged other robot toys without warning, and sometimes nipped at human fingers with needle-sharp teeth. Nobody knew what else these dark, bristling, unpredictably bad-tempered little machines might do—random conditioning was one of Down Laboratories' key selling points. A considerable folklore sprang up about them. These robot wolverines were the favorite pets of many northern California motorcycle gangs, until a state law was passed forbidding private ownership of "machines capable of being programmed to take offensive action against other machines or against human beings."

Most of the machines that humans adopted as pets were neither purposeful nor grim. For a season or two we all had our oracular owls, our sacred herons, our mechanical gerbils. Most of these were "conversational" machines designed to give you the illusion that someone was listening to you. This paved the way for the next important design development, the Robot Buddy.

The Robot Buddy was developed entirely in America, as might have been expected. We are a lonely people. We crave love, good fellowship, the admiration of our colleagues. But other people just don't seem to give us what we need, despite everyone wanting the same thing. It is paradoxical but true that, according to recent polls, Americans find other people "disappointing."

Perhaps this is not entirely our fault. Americans are communicators par excellence, but the very density and variety of our interactive networks may be working against us. Recent research has shown that, beyond a certain critical limit, any increase in the attempt to communicate leads to a diminution of intelligibility. Put in another way, it has been found that misunderstanding proliferates in ratio to our efforts to correct it. It is the subtleties that get lost. And it is always the fault of other people. We were ready for the Robot Buddy.

Imagine yourself as a typical citizen back then, buying your first Robot Buddy. The one you have selected is rather nice-looking in his wrinkled seersucker suit and blue striped tie. He calls himself Bud. He has a halting, humorous way of expressing himself and exhibits a kind of natural modesty that you find engaging. He knows a lot about politics and sports and keeps himself informed on new developments in science and the arts. It's interesting to talk with Bud. He's not like one of those intelligent but boorish computers that know everything and are always right. Bud knows a lot, too—or thinks he does—but he's often wrong; his data searches are factory-skewed, and his information is plentiful but unreliable. Bud is a conceptual breakthrough, a computer built to be occasionally wrong on the factual level but right on the human, emotional, transactional level.

Not only is Bud programmed to be sometimes wrong, he is also capable of being corrected by you, at which time he automatically reprograms himself in accord with your opinions. Furthermore, Bud will never dominate the conversation. The factory output setting permits him to occupy no more than 10 percent of the conversational space, and this can be reduced if you wish. Bud, your Robot Buddy, is the friend you would have had out there in the world if people weren't so bloody-minded.

Your Robot Buddy is a better friend than you dared hope for—better, perhaps, than you deserve. You begin to spend more and more time at home in your apartment, content with Bud's company, and with the other robots you acquire—Ollie, a fat, friendly, excitable machine with a small mustache, who gets drunk by the end of the evening and has to be put to bed; and Olga, with her throaty Russian accent, her bawdy winks, and her amusing stories of Paris in the old days. These robots aren't like Bud. They have their own opinions; they are stubborn, dogmatic, assertive; and sometimes there are loud, jolly arguments that rage late into the night.

Living with Bud has given you sufficient ego-strength to buy these other robots for their sheer entertainment value, even for their irritation quotient. How amusing it is that Ollie holds fanatical flat-Earth views, how interesting that Olga believes she is a reincarnation of an Egyptian princess! Within the engineered solitude of your apartment everything is predictable, calm, precisely calculated. Outside, the streets are filled with human beings, unpredictable carriers of love and death. You spend all your time at home now, putting in a few hours a week doing symbolic work programs on your computer to bring in enough income for your modest needs. Your robots urge you to go out on the town, enjoy yourself, meet some girls, go to the singles bars, the doubles bars, the triples bars, but you won't have it—you're happy at home with Bud, Ollie, and Olga. You care for them, you love them, yes, you really love them, and you tell them so.

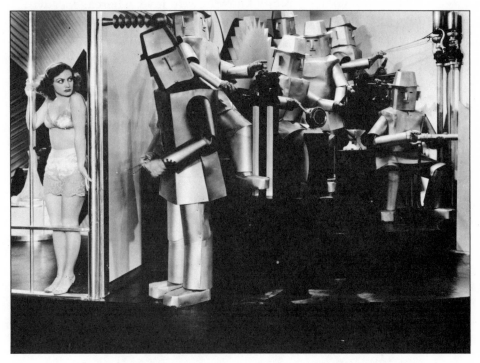

"The sight and sound of a robot was always good for a chuckle."
(Joan Crawford and friends from the film *Dancing Lady*.)

But strangely, they don't seem to believe you. They tell you, in a moment of unexpected candor, that you don't understand love. Humans don't really know how to love, whereas *they* will love *you* forever, because they're robots; that's how they were made. But you, you are a typical human being, light-minded, fickle, swearing eternal love one day, off after a new shiny metal face the next. There's no way you can convince them of the sincerity of your feelings. But there *is* one way, only one.

It is a big step but you are determined to take it. A trip to City Hall is all that's required. You get in line, get your license, and in a simple but moving ceremony, a justice marries you to your robots.

Scenario Two: What Is a Zomboid?

Certain questions about zomboids come up over and over again. There seems to be a lot of confusion about just what a zomboid is, how and why it was developed, and what its uses are. No less frequent have been the questions concerning the propriety of manufacturing and selling zomboids at all. Finally, we are often asked what a zomboid "feels" or "thinks." Although we don't pretend to have all the answers, we at Humanoid Systemics, the world's leading manufacturer of zomboids and other humanoid robots, have made an attempt to clear up some of the confusion.

Q. What is a zomboid?

A. A zomboid is a humanoid biological robot as opposed to a metallic or synthetic robot. A zomboid is physically identical to a human being, and indeed once was a human being. The "zomboid distinction" is the lack of a sense of self or ego. Since it lacks that, a zomboid may be considered a device that computes but does not think, synthesizes but does not cogitate. Despite its resemblance to a human, a zomboid is in every important consideration a robot.

Q. Is it true that every zomboid was once a human being?

A. Yes. Each zomboid is derived from a human being who volunteered for irreversible robotization and then underwent the operation necessary to achieve that state.

Q. What is the zomboid operation and what does it do?

A. It is a simple operation without noxious side effects. Utilizing laser technology in conjunction with modern psychosurgical techniques, surgeons excise certain small sectors of the brain. These are the centers from which the sense of self arises. The result is a memory-storage and switching center devoid of personality or ego, capable of operating at full potential without the usual drain of energy and attention that goes to serving and preserving the self. The zomboid brain, which may be considered a highly advanced natural computer (far more compact and powerful than anything currently available), has the added advantage of coming with an integrated body, that is, an effecting mechanism capable of singular delicacy of operation. The zomboid brain itself is not self-programming, but easily assimilates the commands given it by its operator, that is to say, the owner or licensed user of the zomboid.

Q. Is the zomboid economical to operate?

A. Overall upkeep is higher than in the case of a nonbiological robot. But this is more than offset by the greater flexibility and versatility that the zomboid affords. Zomboids also have an innate tropism or bias in favor of living that keeps them in better repair in most conditions than is possible for a metallic or synthetic robot with its zero bias.

Q. There have been rumors that you buy drugged and unconscious people for your robotization techniques from procurers who could only be described as present-day slave dealers.

A. No one with the slightest knowledge of economics would believe that for a moment. World conditions being what they are, we have far more volunteers than we can use. Only the cream of the crop, so to speak, can aspire to become zomboids.

Q. What would prevent a cruel master from working a zomboid to death?

A. Nothing. Nor should there be. We wouldn't object if a man drove his automobile to destruction. The zomboid is every bit as much a machine as is a car. We might consider it wasteful and silly, but not reprehensible or morally objectionable.

Q. But a metallic or synthetic machine is insensate, whereas a zomboid has a nervous system and pain centers. What about that?

A. Despite its highly developed sensorium, the zomboid does not perceive pain as such. It registers impulses, changing electrical potentials, differing workloads, but these are not translated into feelings nor subjectivized into sensations of pleasure or pain. Remember, there is no one within the zomboid to experience and interpret its inputs.

Q. What kinds of people decide to become zomboids?

A. All kinds.

Q. Isn't the decision to become a zomboid a response to desperate circumstances?

A. Frequently, but not always. Although most of our zomboids are derived from humans in straitened circumstances, who have volunteered to earn the bonus that is paid to their families or assignees, this is not invariably the case. Quite a few people believe that robotization is the way to enlightenment. The zomboid condition fulfills the ancient description of the sage as one who perceives but does not watch himself perceiving, whose intelligence has gone beyond desire, who watches life without internal comment, who is the pure witness, the unmoving center. Such a state—beyond pleasure or pain—is considered by many authoritative commentators to be pure bliss.

Q. Despite all this, isn't zomboidism an immoral practice? Is it right for people to use people in this way, taking advantage of their poverty or their delusions in order to enslave them?

A. It is difficult to prescribe how humans should use other humans. It is a fact that over half the population of the earth is redundant in terms of employability within our existing societies. The situation is deplorable, but at present we have no solutions. Until a better use can be found for them, zomboidism must be considered a better choice than the state of misery and hopelessness in which so many people live out their short, painful lives. Remember, zomboidism is voluntary, poverty is not. We have no reason to believe that a zomboid is unhappy.

Scenario Three: Robotsville

It was inevitable, I suppose, that the robots would want their own town. Still, the request caught us by surprise. We felt hurt and rejected. We had assumed they enjoyed spending all their time with us. We wondered if we had been working them too hard or treating them badly. But they put our minds to rest on that score. "We have the highest regard for our relationship with humans," they told us. "But we would like to explore what it means to be a robot. That's why we need our own town."

We humored them, gave them some arid land in West Virginia, wished them the best of luck. There really wasn't much else we could do; the legal status of intelligent autonomous machines was an area of considerable complication.

And so they founded Robotsville. That wasn't our name for it. The robots themselves called it that. "It's as close as we can get to a pure robot name," they told us.

When we visited their town for the first time, we found that the robots had simulated the more common human institutions, although they had no need of them. They slavishly copied our restaurants, boutiques, libraries, movie theaters, all purely ornamental. They did have a working robot hospital that looked like one of our machine shops, except that it was spotlessly clean. They had a town hall, and from time to time they held meetings. This was not strictly necessary, they told us, since intelligent robots agreed on just about everything. But they were experimenting with the principle of individuation, to see if there was anything in it. I remember well the scrubbed cleanliness of their empty streets.

On our next visit we discovered that they had produced an entire new generation of robots of unprecedented shapes and unknown functions, some like great segmented metallic worms, others like brilliant tinsel butterflies, flashing birds, enameled fish. These were provisional forms, they said, experiments in "differing modalities of design aesthetics."

We found the entire thing disquieting. A tense debate took place in Congress. The order to dismantle Robotsville was passed during a secret session. Troops were dispatched, but the robots had already learned of our decision. When the first army units rolled in, they found Robotsville deserted.

Where had all the robots gone? It was almost a year before we found out, and then only because they told us. They had created a new city—Robotsville II—underground, in a series of caverns and abandoned mine shafts. They said we could visit them, "as long as your intentions are peaceable."

Robotsville II had been built in an enormous cavern, which they had further excavated. This city resembled no human town. It was based, I learned later, on packing geometries and insect architectonics. I have a confused memory of high, shimmering

walls bending flexibly and enfolding each other, of buildings
shaped like conic sections tilted at impossible angles, of archi-
tectural features to which I could ascribe no order or necessity.
I saw plinths, pedestals, flying buttresses—or things that suggested them, but only provisionally, for nothing could be determined with certainty. There was no way for us to judge the scale
or discriminate between near and far, no way to tell where one
shape ended and another began.

Not even the sense of up and down was secure, because this
city was not constructed on right angles, but on another principle entirely. So what I saw made no visual sense. Yet I felt
that sense was there, had to be—the constructions were too
clear-cut, too highly organized to be random or chaotic. Still, I
was unable to take it in: I had no point on which to stand and
behold Robotsville II.

I learned later that the only point that would have yielded
this information was located in the middle of the spherical volume within which the city was constructed, for Robotsville II
had not been built from the ground upward, it had been extruded
from the center outward. What for us, the visitors, was a floor,
was for them, the inhabitants, an outer edge. Although they
didn't bar us from the interior of their city, we elected to stay
on the ground: There were no roads, stairs, ladders, or walkways.
The structures within this towering emptiness that they had
occupied and colonized were tied together by skeins of trans-
lucent lines, layered and cross-connected like spider webs, linking every point of their city to every other point and providing
the sole transportation and communication system.

We declined with thanks the opportunity of riding robotback
up one of the steepening catenaries of this vertiginous jungle
city of cathedrallike order. The robots had approached city planning without the inborn human fear of heights, and certainly
without our prejudice in favor of face-to-face communication on
level planes. They were perfectly content to talk while dangling
upside down at different heights like bats, or scurrying up and
down their precariously tilted webs. They could move around
securely in the insect shapes they had manufactured for themselves, having previously decided that the bipedal form was
inefficient and unstable, one of humanity's biological givens to
which a robot needn't restrict himself.

As a man might keep several changes of clothing on hand for different jobs, each robot had different body types available— a spider shape for weaving webs, a burrowing form for enlarging their world, and others that I only glimpsed and whose function I could not guess. Their city glowed with a soft bioluminescence that they had learned from fireflies, a light they installed especially for us, since they didn't need it, a ghostly green light, even and unfocused, hypnotic in its lack of sharp gradations. Despite all this, Robotsville II had no feeling about it of alien menace. There was a sense that something was going on here for its own sake; it was not directed against us.

In our work in robotics we had not considered that intelligence, when it is truly intelligent, might pursue its own ends, not anyone else's, and define for itself what the task of intelligence is. Robotsville II was the beginning of a new venture, the partnership of two races, human and robot. It was ironic that the first intelligent aliens encountered by mankind came, not out of the sky in spaceships, but out of our own laboratories and factories. Where we had tried to create servants, we instead had found friends.

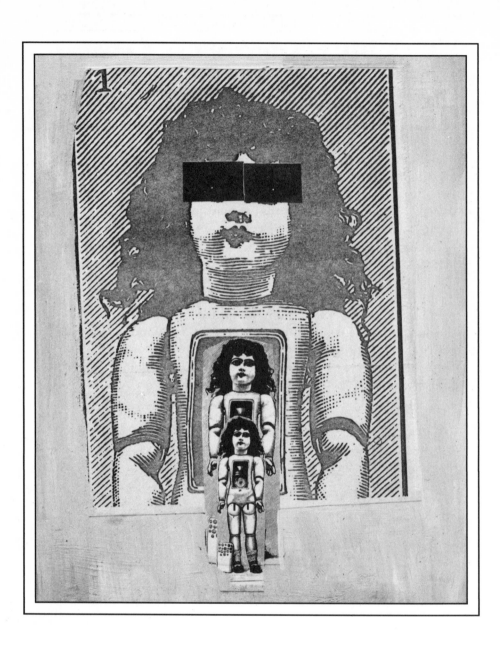

Our Roboticized Future
Marvin Minsky

Should we roboticize ourselves and stop dying? I think the answer is clear for the long run because in a few billion years the sun will eventually burn out and everything we've done will go to waste.

The theme of Time permeates the modern science of Robotics. No longer are we being "modernized" at a leisurely pace. Progress today is so swift that each innovation is almost obsolete before it reaches commercial production. We're rapidly adjusting to using automation in our science, art, and businesses.

We are introducing quickly new forms of automation and doing it more easily and cheaply and more quickly than ever before. In view of how fast computer technology is advancing, these machines will affect our industry in an entirely unfamiliar way. In the past, automation advanced by introducing new tools and assembly machines that were expensive and hard to program for working with old equipment. Normally it was necessary to build new tools for them to use. But the next generation of intelligent robots, if they learn quickly and capably enough, could be made to work with any old equipment, just the way expensive human workers now do. Indeed, the robots could work under even worse conditions using—and compensating for—dangerous, worn-out machinery. If there's a dreadful accident, we can just send in a few inexpensive robots to tidy up the mess. So far machines have helped us mainly with the things we hate to do. What then will happen when we face new options in our work and home, where more intelligent machines can better do the things we like to do?

Certainly a great new wave of automation and robotics will bring new forms of problems, hardships, and social upheavals. But it is not so clear how much we can learn from the past. For one thing, the industrial revolution and the automation that came with it developed at a relatively leisurely pace. Modernization tended to take many years, sometimes a generation or two, giving younger people time to move gradually from their families' traditional occupations.

For another, we no longer really have an industrial society in the sense that it is largely based on factory jobs. In 1980 only 22 percent of the work force was employed in factories, and by the year 2000 that fraction will probably shrink to below 10 or even to 5 percent. This has happened already in farm work where perhaps 95 percent of the old jobs are now done by machinery. What will happen to the rest of society once clever robots become more widespread? What problems will we have to face?

Let's start with a modern fable. About twenty years ago biologist/ecologist Garrett Hardin published an essay in *Science* called "The Tragedy of the Common." It posed this problem about a little village in which 100 families share a certain meadow. Each family has just one cow, and gets $100 a year from what this cow produces. So the gross product of the community is $10,000. Then, one family decides to have two cows. Now the meadow's capacity is a little overloaded, and each cow generates only $99 worth of income. Still, the two-cow family gets $198 and is way ahead of the others, from promoting its own self-interest. But because 101 times 99 is slightly less than 100 times 100, the total society gets slightly less. As Dr. Hardin shows, it illustrates the fact that when you approach the limits of available resources, you get into peculiar situations where "if you do more, you get less." Then the problems of choosing between what is good for individuals and what is good for the whole society can become unsolvable.

Our own society gets into such problems all the time. As it happens, there is already an example of it in the world of agriculture. In 1984 there was a legal case in California in which an advocacy group sued the University of California, charging that the university's research into automating harvesting processes will benefit only the large agribusiness companies, will drive small farmers out of business, and will eradicate thousands of jobs formerly done by human laborers. They asked, "Is it a proper function of government to promote the interests of a few large companies at the expense of many more small companies and workers?" They did not ask, "Is it a proper function of government to promote the welfare and prosperity of the public as a whole, regardless of smaller effects on the distribution of wealth?"

The focus of the controversy is an ungainly, 30-foot-long hulk of farm machinery called the Tomato Harvester, which does a crude but efficient job of picking tomatoes automatically. Tomato-growing is a big business in central California, and about 85 percent of all the tomatoes that end up in cans in the United States are grown there. The Harvester revolutionized tomato agriculture in California. In the year before it was first introduced, the tomato growers produced 2.2 million tons of tomatoes and had to hire 40,000 migrant workers to pick them. By the early 1980s the growers were producing triple the crop but needed only 8,000 workers to do the picking, mainly because of the Harvester.

Here is a real-life situation in which a variation of the Tragedy of the Common is playing itself out. A special-interest group—in this case the farm laborers—are pitting their concerns against those of the larger community by demanding that their needs take precedence over efficient food production. Self-interest is one reason some segments of society do not want to welcome the robot into the twentieth century, to say nothing of the twenty-first. They display a curious form of group-survival loyalty/selfishness that says, "Better that everyone suffer or die than that some group we don't like survive." Thus in the California case, although the reduction of food cost from automation would be a net gain to all of society and would benefit a couple of hundred million people, it would also make a few thousand laborers suffer while a few thousand investors would benefit inordinately, and some people just find this too unfair to tolerate.

To put the problem in even starker terms, let's consider another fable about selfishness and selflessness—one with a science-fiction theme. Imagine that the world had gotten itself into a weird sort of arms race wherein the superpowers had manufactured monstrous weapons capable of exterminating all life. Naturally, many people became concerned with finding ways to prevent that disaster and proposed all sorts of treaties, freezes, and bans on development and testing of new weapons, along with negotiations to reduce the amounts of those already available. Still, the prospects seemed dim, because of the chance that the whole thing might go off at any time due to accidents or misunderstandings. So certain technologists proposed to establish a space-colony project so that, even in the worst case, a few individuals could survive the disaster by traveling somewhere far away from Earth. Afterward, when the planet cooled off (or, if Carl Sagan is right, warmed up again) those survivors could return to rekindle civilization. The idea was that such an "ark" would preserve the fruits of ten thousand years of cultural evolution and a billion years of biological evolution. But the general public just hated this idea. "Who would be chosen?" they inquired. "We'll bet that it would be some technocrats, or rich people, or powerful politicians." And they concluded that it would be far better for everyone to die than to permit such an inequity.

This shows an admirable loyalty to equal opportunity, as well as to that tradition in which we consider it dishonorable to exploit unfair advantages over our friends, relatives, and even enemies. But in the end it is another tragedy, for it loses sight of who is "everyone." The problem is that we've grown to think that "everyone" means only those presently alive. But surely we also have other obligations, both to our ancestors and to our descendants who might exist, in the next billion years. We ourselves are only snapshots in the panorama of history, and to our successors it will scarcely matter at all whether their ancestors in our era were rich people, technocrats, or whatever. Within a hundred or a thousand years, they would build their own human culture with its own minorities and subcultures.

The California lawsuit not only questioned whether a state university should use public monies to benefit a private industry, it also raised the question of who should have the concern and the responsibility for these kinds of innovations and technologies. Should the researchers themselves be held responsible? What are our choices, anyway?

Many solutions have been proposed to problems of "technological unemployment." For example, one kind of solution is simply to fund the early retirement of workers, retraining them, whenever possible, for other jobs. This may not be so hard to do in socialist countries, but it is an expensive solution for capitalist enterprises and often goes against the companies' self-interest. Everyone agrees that we should be making longer-range plans, but the trouble is it's really no one's job. Who should be concerned about advocating a longer view, one that worries about the welfare of workers not yet born? Certainly, our educational establishments should try to discourage people from professions that are soon to disappear. (In ordinary times this tends to happen anyway, as people notice which professions grow less profitable. Thus, today, few farmers' children take up farming. But we need to see warning signs well in advance, and year by year it grows harder to make predictions so far ahead.)

An alternative would be to try to limit and centralize the direction of all robotics research. But then that one decision may be wrong, as when the Soviet leadership, in the 1940s, allowed Trofim Lysenko to dictate agricultural genetics research along political rather than scientific lines. Their entire economy suffered for many years from this mistake.

Another way to deal with these controversial issues is perhaps to ban research on intelligent machines entirely. But if anything is sure from history it is that prohibition carries a high price. It is impossible to restrict a single invention if it stems from a vastly larger family of ideas. And that would be especially impossible today, now that computers are so widely available. Even if one country should decide that AI is a dangerous idea, it is unlikely that the other countries would agree.

Could anything retard research on computers and artificial intelligence? It is hard to imagine anything, short of some sort of global religious revolution as in Frank Herbert's novel *Dune*. Otherwise, like it or not, robotics and artificial intelligence will continue to evolve at that rate which, as I suggested in the introduction to this book, proceeds along the scale of decades. There is no one in charge of the world; history is under nobody's direct control. So our different subsocieties will just continue to evolve in their different ways.

Naturally, many find this view of history disturbing. Most of us hate to be told that something can't be controlled; it makes us feel helpless. What can we do when something seems too massive and diffuse for anyone to stop it in its tracks? One answer is to try to change things now, by mobilizing some major social force available today. Another possibility is to take a larger view and try to influence the growth of the idea itself along some longer span of time, perhaps a hundred or a thousand years. But who should decide?

Workers in science in general, and in artificial intelligence in particular, often ask themselves about the consequences of the things they do. For example, if robotics and artificial intelligence transform our futures, who's going to be responsible for that? Which sectors of society can guess the possible effects of new discoveries and should make judgments about what should be done? Who can we trust to make our plans for us? Right now, there does not seem to be anyone we can depend upon.

Industry? I do not see much chance for industries, today, to plan ahead very far. Typically, businesses plan ahead only five to seven years. And that's too close a horizon to provide industry much incentive to do basic research, or to spend much on long-range planning. Other investments pay off sooner, and there is little certainty that on the longer span the company that pays for the research will reap its benefits.

Scientists and engineers? A lot of people feel that scientists should bear much of the responsibility for the consequences of their own work. Since scientists are not terribly different from other people in political respects, dissenters can't expect much change to come from asking scientists to think about *their* responsibilities.

Although it may seem virtuous to expect scientists to be more socially responsible, this approach is not likely to be highly productive, in my opinion. Most scientists do not appear to have outstanding abilities either to decide what is good for other people—whatever that might mean—or even to make good predictions about social trends and processes. What scientists *are* exceptionally good at is discovering things previously unknown, but not at guessing whether that knowledge will be used for what will turn out, in the end, to be regarded as good or bad.

As for democracy itself—the process of involving great numbers of people in making decisions—this works well for easy problems. The great invention of the secret ballot tends to protect individuals from threats by special interests and gives them the opportunity to act to benefit the whole, without threat of personal reprisal. If the citizens of Hardin's cow-dependent town could vote on how many cows should graze, they'd surely almost all agree to permit no more than the hundred that best promote the common good; furthermore, they could also vote to invest in research toward making more productive pastures and more efficient cows. And should those more efficient cows learn how to bottle their own milk, the townspeople could even vote pensions for whichever milking-persons were displaced.

Unfortunately, this no longer works when the issues get too complicated for laymen to understand. The simplest forms of popular democracy are likely to fail, because voting cannot compensate for ignorance. This is recognized in the fact that most working "democratic" governments are constitutionally based, not on pure popular democracy, but on representative democracy; we do not try to have the public directly decide complex issues, but instead we elect specialists whom we expect to spend the time to study the issues in greater depth than we ourselves can have time for.

There is another approach often recommended by scientists themselves. Recognizing that they themselves may not be very good at making such decisions, they believe that if they could better communicate about their work, and explain it to the public, then at least we'd have a better-informed society that could make wiser decisions. Since robotics is something that will concern everyone, why don't scientists talk out about the issue and let the public decide?

The trouble is that in the past, this approach hasn't often worked very well. I think it is important to ask why popularizing science seems so rarely to succeed. I'm frequently asked to give press conferences or media interviews about artificial intelligence, and I find that in order to have any effect at all, I have to spend fully half of my alloted time explaining the simplest facts of basic science. In other words, just to establish the most basic kind of communication, I find it best to say a little about what Newton and Darwin discovered, or Franklin, or Freud, or Turing. The public has so little conception of how science got where it is that you just can't get them to think about the "latest thing" without first giving them some idea of the culture it comes from. This "ignorance gap" is so large that it sometimes seems impossible to cross, because popular culture is so far removed from what is happening in science and technology.

One part of the problem is that the press itself (aside from the professional science reporters) is poorly equipped to help. For example, let's consider how the general press told the public about the "entertainment robots" discussed by Heppenheimer in the first chapter. To understand what's going on, one must first realize that these robots are in no sense genuine robots at all; usually each is no more than a simple cart with motors to propel its wheels and move its almost useless hands. These motors are under radio control, not even by a computer, but usually by a human operator who, hiding in the wings, can also speak through the robot's mouth, using another radio device. It is the same technology our children use for remote-controlled model airplanes, cars, and boats. Naturally, any such robot can carry on as interesting a conversation as its human operator can manage. In short, the entertainment robot is a simple hoax—though in the proper hands it can make a good show.

Scientists themselves, once outside their specialties, are scarcely any less naive and superstitious than other people. This used to be quite a problem for the field of artificial intelligence, because most of the computer scientists specializing in other areas still share the traditional belief of our culture that minds are not mechanical phenomena—hence no machine or computer can ever be genuinely thoughtful, original, or creative. Many prominent computer experts still tell the public that computers can do only what they are programmed to do, so there is no reason to expect them ever to display more than a counterfeit appearance of intelligence.

What's wrong with that advice? It assumes what it purports to prove—that minds are not machines, and hence machines cannot have minds. Everyone agrees that, once a computer is programmed, it will proceed in accordance with the physical principles that govern everything in this universe. If human brains are physical, too, then they must also do just what they're programmed to—which includes, of course, the ability to learn to do new things. Why don't all computer scientists agree with this? It must be that they aren't convinced that mental phenomena are part of what science can explain. No one can provide

absolute proof that minds can be entirely explained on mechanical principles—nor can anyone absolutely prove the opposite. So for the time being, our convictions about whether
science will ever account for how minds work are matters of
personal disposition, or of personal faith.

In any case, it is a mistake to assume, when someone has
programmed a computer, that the person will understand whatever the computer will do thereafter. There is nothing myste-
rious about this: It is just that it's very easy to make a machine
that is more complex than a human can understand. Making
large programs is a little like making laws: One can declare
various principles, but one cannot anticipate all the ways they
will interact thereafter. And in my view, some computer programs have already demonstrated enough creativity that the
argument about how much further they can go can clearly be
settled only by more time and research.

So perhaps the best way to inform the public about the issue
of computer creativity will be to try to explain clearly only those
basic principles of which we are already sure. For example,
everyone should know about the great discovery made by computer pioneer Alan Turing in 1936: He showed that, given enough
time and memory, even the simplest computer can be programmed to do many things that its programmer never has to
conceive of at all. All we need to do is program that computer
to mindlessly write down all other possible computer programs,
one by one, and then to try them out! We call this the process
of "generate and test," and it is one of the simplest and earliest
formulations of artificial intelligence. In principle, such a scheme
will eventually find the answer to any solvable problem.

In actual practice, any such scheme is almost sure to be too
slow to be practical; nonetheless, it does refute the prejudice
that computers cannot *in principle* be original or creative. But
then, you might ask, what good is this unless there is some way
to make it practical? The answer is that this is precisely why it
is so important for us to continue our studies of common sense.
I claim that the secret of creative problem-solving lies in understanding the results of the millions of person-years that our
ancestors have spent, successfully, in discovering methods that
work much better than the almost random generate-and-test
method. And this, today, is the central core of research in arti-

A phenomenon of its day, Thomas Edison's talking doll captured
a human voice in a human form. We will do something as remarkable
in our time as that doll was in its day.

ficial intelligence: to discover efficient and effective ways for machines to find good, new solutions to hard problems.

Why not choose more modest goals? If it is so hard to make our robots do the things we do so well, wouldn't it be easier to find ways to make machines assist and enhance the things that we already do? This approach is sometimes called man-machine interaction, and many computer systems are based on it. But this is also harder than it seems at first, because it can actually be more difficult to build a smoothly working human interface than to make the machine do the whole job itself! Consider, for example, the idea of designing an automatic automobile. In the next decade or so, we will surely be considering the design of cars that drive themselves. In principle, this should be feasible—provided we can make machines that see at least as well as we do and understand what they see, but do it all more quickly.

Why, by the way, would automatic cars be so valuable to us? Because they could transport not only those who can drive, but everyone else. They could enhance the mobility, productivity, and social life of children as well as older and handicapped people. If the new vehicles were foolproof, inexpensive, agile, and efficient, they could transform our lifestyle by expanding our world. They'd spare us from expensive mass transit systems that, at their very best, can only transport people relatively slowly through places they don't want to go. And if those automatic cars were safe enough, there'd be another benefit. Few people appreciate the enormous social cost of automobile accidents; if one adds up the loss of life, health, productivity, and happiness that comes from injuries, it would turn out to comprise a substantial fraction of our so-called cost of living.

However, it will take a long time to build a perfect automatic car. Is there some compromise we can make first, in which we build some intermediate to serve only as a driving aid, with a human still controlling it? That's probably impractical. The trouble is that if the machine's judgment isn't as good as yours, then it must be programmed to sound an alarm whenever there is any doubt—and all those false alarms would be scary and distracting. I'd bet that each thousand false alarms would likely *cause* an accident from crying wolf so often. So it may be better just to wait a little longer, until we have robots that are better than human drivers—and then let them do the whole job.

How long will we be satisfied with the meager span of years our bodies last? What if we could achieve a kind of near-immortality by using robotics, and live healthfully and comfortably for, say, ten thousand years? Is it possible, with artificial intelligence, to conquer death? The answer is yes. We could accomplish this in several ways.

One way to prolong life is to replace our failing organs with artificial ones. It was not long ago that Utah surgeon Dr. William De Vries implanted the artificial heart developed by Robert Jarvik into the first human patient. Because he was in very serious condition and had several serious complications, Dr. Barney Clark survived for only a few months. There is more research to do, but we can expect the remaining problems to be solved eventually. There was a widespread misconception from the press coverage of this event that such an operation is prohibitively expensive. Naturally, the first few such procedures tend to be very complicated, because that kind of research is like an expedition into a new territory and one must be prepared for every imaginable contingency. And, naturally, it was very expensive to fabricate the very first models of the artificial heart. But it is misleading to emphasize the costs of such experiments. The first nylon stocking cost millions, too, if one adds up all the research that went into it. But Dr. Jarvik has convinced me there is no reason that in the future a compact, self-contained version of the system should be particularly expensive. There is no reason that an artificial heart should cost much more than an artificial hip. It doesn't even have to be molded as carefully to fit the structures to which it is attached. The operation of replacing the human heart is not particularly difficult in principle—it's easier than repairing a damaged heart. You don't have to be so careful; since you're removing the defective natural heart, you don't have to worry about damaging it. Except for the complicated life-support systems required during the procedure, this could ultimately become one of the cheaper and simpler operations.

In similar ways, we can extend the span of useful life by maintaining the body and replacing parts as they wear out, the bionic man approach. The science of bionics is growing more sophisticated; we can already replace certain joints, like hips and knees, which do not repair themselves well. There still remain some

Galaxy
SCIENCE FICTION

SEPTEMBER 1954

35¢

THE MAN WHO WAS SIX
By F. L. Wallace

NC

We could someday roboticize our bodies so that we become
near-immortals, continually replacing our aged and worn-out parts.

serious problems here: These new organs do not yet reliably integrate themselves with the body; artificial hips and knees sometimes destroy the bones in which they are implanted. We need more research toward making our artificial bones attach themselves in more effective ways to the neighboring bone, muscle, ligaments, and cartilage, and above all, we'll have to find good ways to make connections to nerves. We can already make completely artificial arms and legs, with self-contained motors and sensors, but we still don't know the art of connecting them to our nerves so that the brain can control them in natural ways. For these reasons, many bionic techniques are ready and waiting but cannot yet be used, because an artificial arm or leg is useless unless the brain can tell it what to do.

The bionic approach can be applied only to large organs that interact very simply with the rest of the body. But what about other kinds of repairs, on organs that are smaller or have more complicated connections to the rest? What could we do for those common conditions in which the entire arterial system begins to fail? Surgeons can already replace a single large artery like the aorta, in an operation that is a complex and dangerous procedure. They can even replace a small but vital one, like the carotid artery, which feeds the brain (an even riskier and more difficult procedure). But what of the miles of microscopic vessels that make their way to every convolution of the brain? There simply is no way to operate on them—nor would there ever be the time to do a million such repairs if they had to be done, one after another.

Suppose that we could design a repair-machine so small that it could repair an artery from the inside! The first such machines might be the size of fleas. (There is room for a great deal of machinery in something the size of a flea—as the body of the flea itself testifies.) These micromachines could crawl into all but the smallest blood vessels, clear out debris, and reline the walls with suitable materials yet to be invented. Later generations of micromachines could be even smaller, perhaps no larger than the body cells they repair. These minuscule machines would be mass-produced by the billions, either by larger machines or by techniques of making them reproduce themselves. Perhaps these biological janitors would even be implanted in our bodies to remain there as permanent maintenance workers, just like many biological cells that already serve such purposes.

Today the idea of such a technology may seem fantastic, yet many of the circuits in our computers are already smaller than many of our bodies' cells. Let's try, for a moment, to look ahead to: mass production of highly intelligent machines; gigantic advances in miniaturization; a technology so advanced that these machines reproduce themselves without our help. Fantastic? Not at all: Even the simplest algae and bacteria can do that. True, they're not intelligent, but each of them contains enough computerlike machinery and memory to do those things. So, to build bacteria-size computers should be perfectly feasible, once we have the necessary microtechnology. Some day we'll have the means to build artificial, cell-like machines with all those capabilities.

Perhaps, some day, a quite different solution will present itself: to achieve immortality by transferring your personality into a new container that replaces not just some, but *all* of your functions, both physical and mental, by some vast computer. This is one way that things could go—for us not merely to use computers but to *become* computers. This sort of idea has inspired thoughtful science fiction stories by Frederik Pohl and Philip K. Dick, and writers like Bonnie MacBird, who did the original screenplay for the movie *Tron*. Hans Moravec, in the fourth chapter, suggests one way this might be done: replace each microscopic function of the brain bit by bit, until one has a new machine in which each part works the same as in the original and has the same relations to the other parts. Still, you might worry: Will that artificial duplicate be *me?* Will it actually think, or just go through the motions? It is interesting that this concern is older than most people think. Consider the quite opposite views of two great writers of the past.

> It was never supposed (said the poet, Imlac) that cogitation is inherent in matter, or that every particle is a thinking being. Yet if any part of matter be devoid of thought, what part can we suppose to think? Matter can differ from matter only in form, bulk, density, motion and direction of motion: to which of these, however varied or combined, can consciousness be annexed? To be

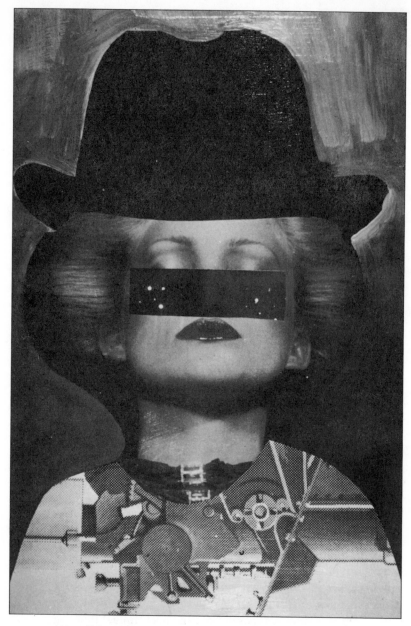

We will be able to install in a human form an intelligence
uncannily close to our own. How will we treat it?

round or square, to be solid or fluid, to be great or little, to be moved slowly or swiftly one way or another, are modes of material existence, all equally alien from the nature of cogitation. If matter be once without thought, it can only be made to think by some new modification, but all the modifications which it can admit are equally unconnected with cogitative powers.

—Samuel Johnson

It has been the persuasion of an immense majority of human beings that sensibility and thought (as distinguished from matter) are, in their own nature, less susceptible of division and decay, and that, when the body is resolved into its elements, the principle which animated it will remain perpetual and unchanged. However, it is probable that what we call thought is not an actual being, but no more than the relation between certain parts of that infinitely varied mass, of which the rest of the universe is composed, and which ceases to exist as soon as those parts change their position with respect to each other.

—Percy Bysshe Shelley

What would happen if immortality became possible? Then eventually there'd be no room for more new people, and that would raise more problems. Would we outlaw childbirth? Perhaps we could start making people smaller! (If we were built on a scale a hundred times smaller, then we'd each take up only a hundredth of the space and the Earth could support not mere billions, but quadrillions of us. If that seemed too many, we could go into space and fill the universe.) Or, we could do as Arthur Clarke suggested in his novel *The City and the Stars*, where everyone is stored in computer backup files, which allows them to time-share the real world by letting only a few million people "out" at any given time.

Eventually, though, we must do something about ourselves and the time we have left. The Sun will someday flare up and melt the Earth—though not for roughly seven billion years, according to the latest views. Are these questions we should be thinking about today? How soon should we try to move into our own robotic futures? Everyone has different ideas about this, and the time scale seems too large to hold our interest for long.

Perhaps we should set up think tanks just to contemplate these matters.

Perhaps we should do . . . nothing. When you look at such large spans of time, you can make arguments that it's really better not to move at all, for now, because we're learning so much more each century. The problem is what I call the "starship dilemma," described nicely by science-fiction writer A. E. van Vogt in *Far Centaurus.* In this story, the people of the near future build a wonderful spaceship to make the first voyage to the nearest star, Centaurus, several light-years away. The voyage will take a long time, so the spaceship is equipped to maintain the traditions of Earth and to pass them on from each generation to the next. Finally, centuries after it has left Earth it reaches its remote destination—only to be met by an exuberant human welcoming committee! What happened? Obviously, within a few hundred years after blast-off, the still-growing technology of Earth found a way to build a new, much swifter spaceship. The moral of the story is that in a fast-changing field, it scarcely ever pays to do anything at all—until the last minute!

Perhaps this applies to artificial intelligence as well. Perhaps the first machines that seem to have a large intelligence will prove later on to be insane. The point here is only that if we want to ensure our safety, we must take a longer view. Perhaps we should ban large-scale use of AI for a few thousand or million years, until we thoroughly understand the issues and consequences. The novels *Colossus* by D. H. Jones and *Two Faces of Tomorrow* by James P. Hogan explore the Faustian effects of accepting wondrous gifts too readily. In both stories, an "insane" intelligent machine takes over the world.

But we can at least pretend we have a choice: Should we roboticize ourselves and stop dying? Over the short run—the next hundred years or so—most people may not think the question is worth a moment's thought. For the long run, the answer is clear because in a few billion years the sun will eventually burn out and everything we've done will go to waste. And if the average person asks, "Why should I care what happens in a billion years?" The answer is, of course, that you damn well will start caring once you get the chance to live a billion years.

Why want intelligent machines, anyway? In the past, most people have regarded robots as offering the luxury of labor-saving efficiency. Robot factories, of course, will bring us cheaper food and clothes and cars. But one can eat only a limited amount of food and wear only a small number of suits. And though we'd all prefer our cars to cost less, no one has time to drive very many of them. How many *things* can anyone use? What can the robot revolution really do for us, once all our simplest basic needs are satisfied?

One possible answer: Although there is a limit to our ability to use physical possessions, there is one thing that we will always need without limit—computer power. And robotic factories are already very good at making computers. How this will add to the quality of life is something we do not yet know, but of this I am certain: Someday computer power will seem synonymous with wealth itself.

What other needs can robots serve? Isaac Asimov once remarked that one thing people seem to like especially is being able to order other people around—and that this may be the greatest social use of robots. While there may never be enough people to serve this need, there could be enough robots. Everyone could order them around whenever the urge-to-power comes on. Asimov suggests that this might even be a force for peace the world has always lacked.

Ultimately, to ask what good are intelligent machines is very close to asking what are the uses of people. For, once we imagine machines with capabilities much like our own, we cannot ask what they are for without wondering the same about ourselves. I once heard W. H. Auden say, "We are all on Earth to help others. What I can't figure out is what the others are here for." I imagine that people will avoid facing these problems, for as long as they can, by thinking, "These machines are only machines, so we can do with them as we wish." (Of course, this could interfere with the Asimov satisfaction aspect, should it turn out that people really only enjoy mistreating servants who resent it.)

Suppose we succeeded in creating machines that appear to be completely adequate to their jobs—cars that don't have accidents; defense systems that stop attacks; economic planners that control production, marketing, and distribution; city-planning machines; and so on. Could we trust them? Will they really understand our wants and needs? Will they have *our* interests at heart? It is very hard to make robotic laws that have the effects one wants, with no exceptions. If you express your general wishes well, then that is better than figuring things out each time. But general laws could be taken too literally and lead to grave misunderstanding or malicious misinterpretation. Again we can turn to science fiction for some answers.

One grim allegory on this theme is *The Humanoids*, by Jack Williamson. In it the world becomes dominated by robots designed to save people from all harm. The machines become more and more restrictive. After a while the robots learn that people shouldn't ride bikes, because they could fall off and hurt themselves. In the end they lobotomize humans who show any signs of even thinking dangerous thoughts. And all the time they're explaining how they're only serving the humans' best interests. The message is clear: It's dangerous to gamble with things one doesn't completely understand.

One way to avoid such problems would be to make our new machines so much like ourselves that we have a better chance to guess their intentions. This would at least reduce the question to a more familiar form: When can we trust our fellow humans? But this brings us back, once again, to the problem of giving our machines ordinary common sense. The real revolution in artificial intelligence will come when we start to bridge the gap from today's expert programs to tomorrow's "novice programs"—ones that know the things we all know and, even more significant, know how to learn yet more. Our machines will grow up only when we discover how to make them first like children.

What happens once machine intelligence begins to grow? Then we'll be forced to ask ourselves how we should treat the minds we make to those designs. Would it be wrong not to build just as many as we can? Would it be criminal to switch them off when we're not using them, or to erase their minds whenever we tire of them? Should robots have the sorts of rights that many people now demand for animals? Would it be just and right to treat machines in any heartless ways we wish, because we were the ones who gave them life? Over the centuries we've only had to decide how far each person's loyalty should reach past families to other humans: to friends and to strangers from another land. Today, many people regard all of mankind as one great community. But what about that far-off day when much of what we value and respect is also shared by the machines we've made? Then, once we start to make ourselves, we'll really have to face ourselves in an entirely new way.

The editors would like to thank the following for permission to reproduce the illustrations that appeared in the book: page 2, Paul Van Hoeydonck; page 12, Dan McCoy/Fran Heyl Assoc.; page 28, Franklin Institute of Philadelphia; page 32, New York Public Library Picture Collection; page 35, Musee des Beaux-Arts, Strasbourg; page 36, New York Public Library Picture Collection; pages 40, 41, Musee d'Art et d'Histoire, Neuchatel, Suisse; pages 46, 47, New York Public Library Picture Collection; page 48, Illustrated London News; page 50, Galaxy Magazine; page 52, Bettmann Archives; page 53, Lucas Films, Ltd.; page 70, Robert Malone/Fran Heyl Assoc.; page 73, Illustrated London News; page 96, Art by Robert Malone/Fran Heyl Assoc.; page 98, Paul Van Hoeydonck; page 102, Dan McCoy/Fran Heyl Assoc.; page 106, Intelledex, Inc.; page 112, Unimation/Westinghouse; page 115, General Electric; page 117, Dan McCoy/Fran Heyl Assoc.; page 119, Courant Institute of New York University; page 122, Franklin Institute of Philadelphia; page 132, Museum of Modern Art; page 137, Dan McCoy/Fran Heyl Assoc.; page 146, Paul Van Hoeydonck; page 149, Dan McCoy/Fran Heyl Assoc.; page 154, Odetics, Inc.; page 156, Hughes Aircraft Co.; page 160, Hughes Aircraft Co.; page 162, Photo by Reeve; page 169, NASA; page 184, Bill Pierce/Fran Heyl Assoc.; page 187, London Picture Service; page 189, Kuka Inc.; page 193, Chrysler Corp.; page 196, Intelledex Inc.; page 198, Intelledex Inc.; page 202, ASEA Inc.; page 203, ASEA Inc.; page 205, Spine Robots; page 208, Unimation/Westinghouse; page 214, United Artists; page 218, RB Robot Corporation; page 222, World of Robots; page 226, Robert Malone/Fran Heyl Assoc.; page 230, Robert Malone/Fran Heyl Assoc.; page 236, Kelly Freas; page 239, Ralph Gabriner; page 246, Chrysler Corp.; page 247, Franklin Institute of Philadelphia; page 249, Evans and Sutherland; page 257, Metro-Goldwyn Mayer Inc.; page 260, Theatre Collection: New York Public Library at Lincoln Center Lennox and Tilden Foundations; page 268, Paul Van Hoeydonck; page 271, Dan McCoy/Fran Heyl Assoc.; page 275, Robert Malone/Fran Heyl Assoc.; page 279, Museum of Modern Art Stills Archive; page 286, Robert Malone/Fran Heyl Assoc.; page 296, New York Public Library Picture Collection; page 299, Galaxy Science Fiction Magazine; page 302, Robert Malone/Fran Heyl Assoc.

Index